D0824440

OF MICROBES AND LIFE

LES MICROBES ET LA VIE

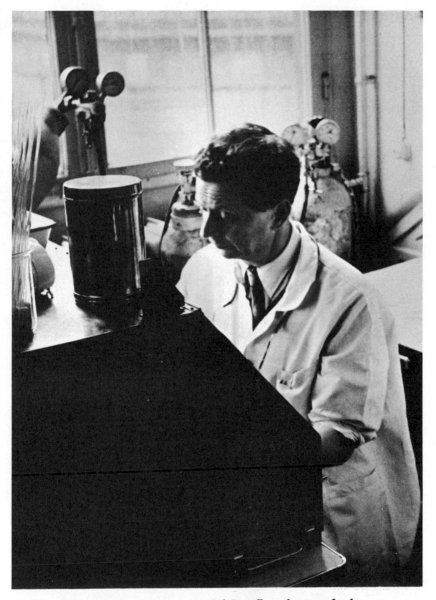

The master craftsman: André Lwoff at the transfer box.

JACQUES MONOD ET ERNEST BOREK
REDACTEURS

LES MICROBES ET LA VIE

1971

COLUMBIA UNIVERSITY PRESS / NEW YORK ET LONDRES

JACQUES MONOD AND ERNEST BOREK
EDITORS

OF MICROBES AND LIFE

WITHDRAWN

1971

COLUMBIA UNIVERSITY PRESS / NEW YORK AND LONDON

Copyright © 1971 Columbia University Press
ISBN: 0-231-03431-8
Library of Congress Catalog Card Number: 71-133382
Printed in the United States of America

Contributors

JACQUES MONOD, *Editor*
Institut Pasteur, Paris

ERNEST BOREK, *Editor*
Department of Microbiology, University of Colorado Medical Center, Denver

L. ALFÖLDI
Institute for Microbiology, University of Szeged, Hungary

ROBERT AUSTRIAN
The John Herr Musser Department of Research Medicine, University of Pennsylvania School of Medicine, Philadelphia

LANE BARKSDALE
Department of Microbiology, New York University School of Medicine

GEORGES COHEN
Institut Pasteur, Paris

SEYMOUR S. COHEN
Department of Therapeutic Research, University of Pennsylvania School of Medicine, Philadelphia

M. DELBRÜCK
Division of Biology, California Institute of Technology, Pasadena

RENATO DULBECCO
The Salk Institute for Biological Studies, La Jolla, California

EMMANUEL FAURÉ-FREMIET
Collège de France, Paris

PAUL FILDES
Melton Court, London

MARC GIRARD
Institut de Recherches sur le Cancer, Villejuif, France

NEAL GROMAN
Department of Microbiology, University of Washington School of Medicine, Seattle

E. F. HARTREE
Agricultural Research Council, Animal Research Station, Cambridge, England

W. E. VAN HEYNINGEN
St. Cross College, Oxford University

S. H. HUTNER
Haskins Laboratories, New York

FRANÇOIS JACOB
Institut Pasteur, Paris

DALE KAISER
Department of Biochemistry, Stanford University School of Medicine, California

NIELS OLE KJELDGAARD
Department of Molecular Biology, University of Aarhus, Denmark

A. KIRN
Institut National de la Santé et de la Recherche Médicale, Institut de Bactériologie, Strasbourg

B. C. J. G. KNIGHT
Department of Microbiology, The University of Reading, England

HILARY KOPROWSKI
The Wistar Institute of Anatomy and Biology, Philadelphia

S. E. LURIA
Department of Biology, Massachusetts Institute of Technology, Cambridge, Massachusetts

J. MILLOT
Musée de l'Homme, Paris

A. M. PAPPENHEIMER, JR.
The Biological Laboratories, Harvard University, Cambridge, Massachusetts

M. R. POLLOCK
Department of Molecular Biology, University of Edinburgh

A. QUERIDO
Academisch Ziekenhuis Dijkzigt, Rotterdam

BERNARD ROIZMAN
Department of Microbiology, The University of Chicago

S. FAZEKAS DE ST. GROTH
CSIRO, Epping, New South Wales, Australia

PIERRE SCHAEFFER
Départment de Microbiologie, Faculté des Sciences, Orsay, France

LUIZ H. PEREIRA DA SILVA
Institut Pasteur, Paris

R. Y. STANIER
Department of Bacteriology, University of California, Berkeley

À André Lwoff

à l'occasion du cinquantième anniversaire

de son entrée en Biologie

To André Lwoff
for the Fiftieth Anniversary
of his immersion in Biology

Préface

Les essais réunis dans le présent volume sont associés avant tout par le lien d'affection et d'admiration que leurs auteurs portent au dédicataire: André Lwoff.

Les amis, collègues et disciples de Lwoff ont choisi pour ce témoignage l'occasion du cinquantième anniversaire de l'entrée d'André Lwoff dans la carrière scientifique.

Au cours de cette longue carrière, Lwoff a ouvert nombre de voies entièrement nouvelles où se sont engagés par la suite d'innombrables chercheurs, témoins de son influence et porteurs de son message, parfois sans le savoir. La diversité même des sujets et des époques sur lesquels portent ces articles reflèteront, mieux qu'une longue notice n'aurait pu le faire, l'étendue, la richesse et la variété des contributions fondamentales apportées par André Lwoff à la science biologique.

Notre espoir est que le lecteur de ces essais, qu'il soit débutant ou chercheur confirmé, partagera avec les auteurs ce rare privilège d'avoir eu commerce avec un grand esprit et un grand créateur.

ERNEST BOREK

SEYMOUR S. COHEN

RENATO DULBECCO

FRANÇOIS JACOB

JACQUES MONOD

MARTIN POLLOCK

PIERRE SCHAEFFER

ÉLIE L. WOLLMAN

Preface

These essays are connected by a single thread: the affection and esteem in which each contributor holds the man to whom the book is dedicated, André Lwoff. His friends, colleagues, and students wrote these essays to mark the fiftieth anniversary of Lwoff's entry into science. During his happily long career Lwoff made seminal contributions to many areas of experimental biology. His insight and experimental skill formed platforms for the launching of the careers of countless investigators and through them Lwoff's influence penetrated into most areas of biology.

It is superfluous to chronicle Lwoff's career and accomplishments in these lines. The essays in the book do it better.

It is hoped that the reader, be he an aspirant or a seasoned investigator, will share through these pages the high privilege each contributor had in meeting a great scientist and a great man.

ERNEST BOREK

SEYMOUR S. COHEN

RENATO DULBECCO

FRANÇOIS JACOB

JACQUES MONOD

MARTIN POLLOCK

PIERRE SCHAEFFER

ÉLIE L. WOLLMAN

Contents

OF MICROBES AND LIFE

LES MICROBES ET LA VIE

Du microbe à l'homme

JACQUES MONOD

André Lwoff est entré dans la carrière scientifique en 1921, à l'âge de 19 ans. Quand le présent volume paraîtra, cette carrière aura couvert cinquante années. Cinquante années au cours desquelles la Biologie pouvait enfin devenir *une* science, alors qu'elle ne constituait auparavant qu'une collection de disciplines étrangères les unes aux autres, entre lesquelles les idées ne circulaient guère, ni les hommes. Le destin, ou le génie d'André (qui saurait distinguer l'un de l'autre?) fut, au contraire, de beaucoup "circuler" et d'apporter à l'unification de la Biologie une contribution majeure.

Pourquoi choisit-on de devenir homme de science, et pourquoi biologiste? De nos jours cela paraît assez clair: "faire de la recherche" semble (à tort) offrir une carrière comme une autre. Et chacun sait que l'âge d'or de la Biologie est venu (seuls quelques esprits chagrins, comme Gunther Stent ou moi-même, dans des moments de dépression, soupçonnent que peut-être cet âge est passé).

Il n'en était pas de même en 1921. La Recherche pouvait être une vocation, pas une carrière. Quant à la Biologie (celle en tout cas qu'on enseignait à la Sorbonne, du temps d'André, comme du mien huit ans plus tard), c'était encore celle du XIXème siècle. Deux disciplines nouvelles étaient nées vers la fin de ce siècle: la microbiologie et la génétique. Ni l'une ni l'autre n'était enseignée à la Sorbonne (ni ne devait l'être jusqu'en 1945). Personne, nulle part, ne soupçonnait que de leur fusion devait naître la Biologie moderne. André, bien sûr, pas plus qu'un autre. Alors pourquoi la Science? Et pourquoi la Biologie? Sans aucun doute l'influence de son père y fut pour beaucoup, qui avait dû, comme une bonne partie de l'intelligentsia russe, fuir l'oppression du régime tsariste. Le père d'André, comme la plupart des progressistes de cette génération, nourrissait une foi positiviste dans les conquêtes conjuguées de la Science et de la société. La Biologie évolutionniste, celle de Lamarck et de Darwin, semblait révéler l'existence dans la nature d'une loi de progrès et constituait l'un des principaux aliments de cette idéologie optimiste et généreuse.

André, qui professe le plus complet dédain pour toutes les idéologies, haussera sans doute les épaules à la suggestion que son choix (comme le mien) a été influencé par des idées philosophico-politiques. Mais l'aversion même d'André pour les discussions idéologiques (autour d'un samovar, à la Russe) révèle qu'il y a été fortement exposé.

Il est bien vrai cependant qu'André n'est pas un idéologue: il est d'abord un artiste et la beauté foisonnante du monde vivant exerçait sur lui un attrait puissant. La beauté, la richesse, l'étrangeté le fascinaient de cet univers des protistes qu'il découvrait, sous le microscope, lorsqu'il faisait ses premières armes à Roscoff, sous la direction plus impériale encore qu'impérieuse d'Édouard Chatton. C'est cette fascination même qui fait les vrais naturalistes. André en avait tous les dons qui lui valaient de devenir, à un âge bien tendre encore, le collaborateur à part entière d'un des plus célèbres zoologistes de l'époque. Cette collaboration devait aboutir à une oeuvre dont bien d'autres se seraient satisfaits pour leur carrière entière: la découverte et la description d'un groupe entier de Ciliés, les Apostomes. Ciliés parasites, et qui plus est, à deux hôtes, ce qui n'avait jamais été observé jusque-là. D'où, pour ces Ciliés, des cycles évolutifs complexes comprenant de multiples formes et de véritables métamorphoses. Été après été, à Roscoff et à Banyuls, Chatton et Lwoff, bientôt rejoints par Marguerite, chassaient le Cilié sur des carapaces de Crabes ou la hampe d'un polype qu'ils disposaient dans de petits cristallisoirs pour suivre leurs métamorphoses de "tomonte" en "tomite" ou de "trophonte" en "phoronte": ils ne cultivaient pas seulement les Ciliés mais, aussi assidûment, le jardin des racines grecques. J'avoue y avoir, à leur contact, pris goût moi aussi.

Les cycles compliqués des Ciliés Apostomes, adaptation parasitaire typique, posaient nécessairement le problème de leur origine, à partir de Ciliés libres, et de l'évolution morphologique et physiologique du groupe. Dès son initiation à la recherche, André devait donc apprendre à penser en termes d'évolution, à ne considérer la forme actuelle que comme le produit d'une histoire, hors quoi elle demeure incompréhensible.

Chatton et Lwoff découvraient bientôt que l'étude détaillée de l'infraciliature des Apostomes et de ses transformations au cours du cycle leur fournissait une des clés de ce problème. Ces observations les conduisaient à la conclusion que chaque cinétosome est le descendant d'un cinétosome préexistant. Il y avait donc là "continuité génétique" d'un organite cytoplasmique. Notion dont André saisissait toute l'importance et qu'il devait reprendre et commenter à plusieurs reprises bien des années plus tard.

Ces grands travaux couvrent une période de près de quinze ans de la vie scientifique d'André, mais n'occupaient en fait au long de ces années que les mois d'été. André poursuivait à Paris des études de Médecine, selon le voeu de son père mais non sans une grande nonchalance. Il voulait travailler au laboratoire, jouer avec des pipettes et des tubes, pourvu qu'il y eut un petit quelque chose de vivant dedans. Chatton le présentait à Félix Mesnil, qui avait débuté comme secrétaire de M. Pasteur avant de devenir l'élève de Laveran et de Metchnikoff. En 1921, à 19 ans, André entrait dans le laboratoire de Mesnil à l'Institut Pasteur. Il a décrit sa surprise à découvrir, lors de sa première visite, que ce laboratoire "provisoire" (fondé en 1907, toujours debout) ouvrait sur une cour d'écurie, pleine de fumier. C'était une grande chance. L'odeur pastorienne du fumier était bien plus favorable à l'épanouissement d'un jeune chercheur que l'atmosphère poussiéreuse, morne et satisfaite des laboratoires sorbonnards de l'époque dans lesquels André aurait pu aussi échouer.

En ce temps, l'Institut Pasteur comptait encore nombre des représentants de la période héroïque de la Microbiologie, pour la plupart anciens collaborateurs de Pasteur, à commencer par l'austère figure de M. Roux, qui dirigeait alors l'Institut. Tout imprégnée qu'elle fût encore de la personnalité de M. Pasteur, la Microbiologie, telle qu'on la pratiquait dans la Maison, s'était éloignée de sa tradition. Dans le monde des microbes qu'il découvrait, Pasteur ne voyait pas seulement les agents des maladies infectieuses, mais les formes les plus élémentaires de la vie, celles par conséquent qui pouvaient en révéler les secrets les plus fondamentaux.

Les prodigieux succès de la bactériologie des maladies infectieuses, de l'épidémiologie et de l'immunologie avaient, en partie, fait oublier cet aspect si important de la véritable tradition pastorienne (qui cependant avait été maintenue par Duclaux, son successeur immédiat). Pasteur n'était pas un "bactériologiste". C'était un biologiste, dans toute l'acception du terme, fasciné par l'étrange logique du monde vivant et habité par le besoin d'en percer le mystère. André aussi, et c'est pourquoi il lui fut donné de retrouver cette tradition et de l'illustrer bientôt de lumineuses découvertes.

Dans le laboratoire de Mesnil, on entretenait divers protistes en culture pure. Mais non des Ciliés, qui en général refusent toute autre nourriture que des bactéries. André se mit en tête de réussir la culture pure d'un Cilié, et y parvint avec *Glaucoma* (*Tetrahymena*) *pyriformis*. C'était beaucoup plus qu'un succès technique: dès lors on pouvait tenter de préciser les exigences nutritionnelles du Cilié et les comparer avec celles d'autres protistes. Déjà habitué à penser en évolutionniste, André comprit tout l'intérêt qu'il y aurait à développer une conception qui mette en relation l'évolution morphologique des Protistes et celle de leurs besoins nutritionnels, de leur "pouvoir de synthèse".

Disposant, dans le laboratoire de Mesnil, de cultures pures des protistes les plus variés, des Chlorophytes et Leucophytes jusqu'aux Ciliés en passant par divers Flagellés, André pouvait, avec les moyens et les définitions de l'époque, préciser leurs besoins nutritionnels et, dirions-nous aujourd'hui, "ne pouvait manquer de voir" que ces besoins (dès lors qu'on les concevait comme exprimant une évolution biochimique partant des formes les plus indépendantes pour aboutir aux plus dépendantes, tels que les flagellés parasites) traduisaient la *perte* graduelle de toute une série de fonctions biosynthétiques.

Il est bien difficile, pour un biologiste contemporain, de comprendre pourquoi cette idée qui lui semble presque évidente *a priori,* paraissait presque scandaleuse en 1932, lorsqu'André en faisait la conclusion essentielle de sa Thèse, et la défendait avec une passion que certains prenaient pour de l'arrogance. Confusion fréquente: "la modestie sied au savant, mais pas aux idées qui l'habitent" écrit un bon auteur. C'est bien vrai, et surtout (mais pas exclusivement) si l'idée est juste: c'est en l'exprimant de la façon la plus tranchante qu'on en éprouve la valeur. Pour le tranchant, le ton d'André ne laissait rien à désirer, dès 1932.

L'idée que l'évolution ait pu impliquer la perte de certaines fonctions biochimiques se heurtait, à l'époque, non pas à d'autres conceptions scientifiques mais, et c'était bien là la difficulté, à l'idéologie "progressiste" implicite pour qui l'ascendance évolutive ne pouvait, ne devait être qu'un enrichissement. L'idée de "perte de fonctions" paraissait blasphématoire, réactionnaire, pessimiste. Mais à un honorable collègue qui lui opposait des arguments de cette farine, André demandait ingénument comment il (le collègue) trouverait le loisir d'accomplir ses importants travaux s'il lui fallait, pour survivre, demeurer tout le jour exposé au soleil dans une solution de nitrate de potassium. Aux yeux de certains les arbres cachent la forêt, c'est vrai. Mais d'autres, perdus dans le maquis idéologique, ne savent même plus voir le buisson des faits qu'ils ont sous le nez. Immunisé sans doute pendant sa jeunesse, André ne courait pas ce risque.

Certes on savait depuis assez longtemps que certains micro-organismes exigent des "facteurs de croissance" dont d'autres se passent entièrement. Mais la *signification* de ce besoin (sans même parler de la nature de ces facteurs) n'était pas comprise. La notion, plus ou moins implicitement régnante, semble avoir été que des exigences nutritionnelles plus complexes trahissent un raffinement fonctionnel, mais non, bien entendu, une dégradation du potentiel biochimique. Idée comparable, en somme, à celle que les Français se font de leur cuisine, dont l'exigeante délicatesse, à nos yeux, exprime le raffinement de notre culture et sa supériorité sur celle d'autres peuples non encore entièrement dégagés de la barbarie, tels les Anglo-Saxons (idée qu'André lui-même considère comme indiscutable, j'en suis témoin).

Certes le terme de "facteurs de croissance" était employé. Mais il recouvrait la notion la plus vague. Les facteurs de croissance étaient des "substances inconnues agissant de façon catalytique". André comprit que, pour prouver la thèse, il fallait identifier un ou plusieurs facteurs de croissance, élucider leur mode d'action. Or il savait que parmi les Flagellés parasites, il en est qui, vivant normalement dans le sang de l'hôte, ne sont cultivables *in vitro* que si le milieu contient du sang. Aux yeux d'André il ne pouvait s'agir que d'une adaptation parasitaire typique; d'une perte de fonction, mais non d'une exigence spécifique

nouvelle. L'identification du facteur exigé par l'un de ces Flagellés parasites (*Crithidia fasciculata*) réussie au cours du séjour d'André et
Marguerite chez Meyerhof devait apporter une confirmation éclatante de
cette conception: le facteur exigé était l'hématine, premier facteur de
croissance à être identifié (1934) et dont la signification fonctionnelle,
évidente et générale, ne permettait pas de douter que l'exigence manifestée par certains flagellés ne fût l'expression d'une perte de fonction
favorisée, tout au moins autorisée, par la vie parasitaire.

Cette dévouverte considérable devait passer presque inaperçue. La
portée en tout cas n'en fut pas comprise, s'agissant apparemment d'un
cas très "spécial", dont l'interprétation, pouvait-on penser, n'était pas
généralisable. André, lui, ne doutait pas qu'elle le fût, et quelques années
plus tard il s'attaquait à l'identification du "facteur V" exigé par une
bactérie parasite, *Hemophilus influenzae*. La chance le servit ("qui ne
sourit qu'aux esprits préparés", disait M. Pasteur). Peu de temps auparavant Warburg avait identifié le coenzyme qui longtemps porta son
nom avant de devenir DPN, puis NAD. Isolé et identifié par André et
Marguerite, le facteur V s'avéra identique au coenzyme de Warburg.
Leur découverte tombait ainsi dans l'un des domaines les plus actifs et
les plus en vue de la biochimie de l'époque (1936). Il fallait bien se
rendre à l'évidence, les yeux s'ouvraient. Quelques-uns il est vrai avec
stupéfaction. Au congrès de microbiologie de 1936, à la suite de la
communication d'André, le grand Kluyver en personne déclarait qu'il
"ne croyait pas aux substances qui font des miracles".

Désormais cependant la voie était tracée, que bien d'autres après
André devaient suivre. Quelques années après, grâce au génie de Beadle,
l'analyse nutritionnelle, fondée sur les idées avancées par André dès
1932, était associée à l'analyse génétique. La génétique biochimique
était née, dont la méthode fondamentale consiste à provoquer par des
mutations des "pertes de fonctions" biosynthétiques, pour révéler ainsi
les exigences essentielles des organismes, tout en identifiant les structures génétiques qui gouvernent ces fonctions. L'âge d'or de la Biologie
Moléculaire s'annonçait.

C'est par des voies nouvelles cependant que la Biologie Moléculaire devait s'emparer d'André. Le sillon qu'il avait tracé, il le voyait
s'épanouir en une vaste percée (future autoroute) où s'affairaient, nombreux, de puissants et honnêtes bulldozers. Il n'était que de les laisser

travailler. Ce n'était plus le style d'André, qui éprouvait le besoin d'explorations nouvelles.

En 1946, nous eûmes la chance André et moi d'assister au mémorable symposium de Cold Spring Harbor où la discipline nouvelle devait prendre corps et âme. Les membres de l'église du phage, déjà constituée, y étaient au complet, avec leurs trois évêques: Max Delbruck, Al Hershey et Salvador Luria. Max en était bien le pape. *Primus inter pares* c'était lui qui définissait la dogmatique. Etait de foi ce que Max acceptait, hérésie ce qu'il niait. Or il niait volontiers, souvent, en fait systématiquement. Entre autres, tous les résultats des écoles antérieures à la fondation de l'Église ou extérieures à elle.

Par exemple, Max comptait pour rien les travaux de Burnet, Gratia, E. Wollman, den Dooren de Jong sur les bactéries lysogènes. Il refusait de croire au phénomène lui-même qu'il attribuait à des contaminations non reconnues. Pour avoir lu certains de ces travaux, et bien connu Eugène Wollman, expérimentateur aussi scrupuleux qu'entêté, nous n'étions André et moi nullement satisfaits des négations péremptoires de Max, qui avivaient au contraire notre intérêt pour cet étrange phénomène. De retour à Paris, nous discutions longuement de l'expérience à faire: c'est-à-dire de celle qui pourrait convaincre Max de son erreur et de la réalité lysogénique.

Convaincre Max: ce n'était pas une petite affaire. André devait y passer de longs mois à pêcher des *Bacillus megaterium* au micromanipulateur, à suivre intégralement leur descendance, soigneusement répartie dans des "guttules", à repérer le bacille qui soudain se dissolvait sans cause apparente, à prouver que celui-là, et celui-là seul, libérait le virus directement hérité, sous forme latente, de ses ancêtres. Heureusement pour lui, André avait à l'époque une charmante assistante avec qui c'était un plaisir que de "pêcher" des bacilles. Max fut convaincu. C'était un résultat important: la lysogénie avait droit de cité dans l'Église. L'étude en était légitimée. Restait à l'entreprendre. Mais que faire? L'arbitraire de la lyse spontanée d'une petite fraction de la population bactérienne agaçait André. D'autres y eussent vu un délicieux prétexte à statistiques pondéreuses. Pas lui. André déteste la statistique autant que Keilin qui disait à Ronald Fisher: "Quand j'ai des résultats susceptibles seulement d'interprétations statistiques, je les jette au panier".

En cultivant jour après jour ses bacilles, André s'aperçut que de temps à autre, et toujours sans cause apparente, une culture se lysait d'un coup, presque entièrement. Lorsqu'on saurait provoquer et gouverner ce phénomène erratique, plus besoin de statistiques, l'instrument serait en mains. Le projet s'avéra hérissé de difficultés déconcertantes. Les cultures se lysaient, mais pas toujours. Souvent c'étaient les témoins qui lysaient à la place des cultures en expérience. Tous les paramètres imaginables semblaient y jouer un rôle: des cations, des anions, l'origine des peptones, "le pH, le rH et le SH" comme disait un de nos honorables collègues pastoriens. André, infatigable avait une idée nouvelle tous les matins, changeait de conditions et de milieu tous les jours, épuisait ses collaborateurs, Louis Siminovitch et Niels Kjelgaard, et produisait une masse de résultats si compliqués que Max, informé, se déclarait "more bewildered than enlightened".

Jusqu'au jour où l'idée vint à André d'exposer une culture, pendant quelques secondes, à la lampe à U.V. que j'employais pour mutagéner des colibacilles. La culture lysa dans la demi-heure suivante. On recommença immédiatement: même résultat; puis le lendemain, le surlendemain et les semaines suivantes. Fini l'erratisme. Le phénomène se reproduisait avec une régularité métronomique. André avait découvert l'induction, exorcisé le spectre statistique, et prouvé du même coup que *toutes* les cellules d'un clone sont porteuses du virus latent. La nouvelle fit grand bruit. Le contrôle de l'induction ouvrait évidemment la voie à l'analyse des systèmes lysogènes. Max répéta immédiatement l'expérience, et dut s'avouer, une fois de plus, convaincu.

Il serait difficile, je crois, d'exagérer l'importance de cette découverte. Il n'est que de considérer les développements ultérieurs qu'elle préparait ou rendait possibles. Le premier en fut la définition par André de la notion de prophage. C'est-à-dire le postulat selon lequel le virus latent présent, on le savait maintenant, dans toutes les cellules d'un clone lysogène, s'y trouve dans un état fonctionnellement et structuralement différent de celui du virus infectieux (du virion). Nul doute que les premières expériences d'André avec les Ciliés parasites n'aient été l'une des sources de l'intuition si claire qu'il eut de cette distinction nécessaire. La notion de prophage ne fut pas acceptée d'emblée. Il fallut la découverte de Hershey pour qu'elle s'imposât tout à fait. Puis vinrent la localisation génétique du prophage, la découverte de l'induction zygotique, l'analyse, par Jacob et Wollman, du mécanisme de l'induction zygotique, la notion d'épisome, la découverte du mécanisme de la répression génétique. Rien de tout cela n'eût été possible sans la découverte de l'induction, et hors la notion de prophage.

Entre temps le grenier que j'avais connu réservé, ponctuel, ordonné, s'était empli à craquer. J'y étais entré, avec ravissement, secouant la poussière sorbonnarde, en 1945. Elie Wollman, Pierre Schaeffer, François Jacob suivaient à quelques années d'intervalle. Notre premier hôte d'outre-atlantique, S. S. Cohen, arrivait en 1947. Le second, Mel Cohn, en 1948. Puis A. M. Torriani, Gunter Stent, Seymour Benzer, A. Pappenheimer, Mike Doudoroff, Germaine (future Stanier), Martin Pollock, Dale Kayser, Dave Hogness et bien d'autres que je ne puis tous nommer. Le grenier craquait. Sa belle ordonnance, réservée, discrète et un peu hautaine, à la manière d'André, faisait place à un bruyant débraillé, auquel Mel contribuait pour une part majeure tout en organisant de gigantesques expériences qu'il menait, je ne sais comment, infailliblement à bien. Il devenait évident qu'il me fallait partir avec mon groupe avant que le grenier n'explosât tout à fait. Ce que je fis, en 1954, non sans mélancolie, encore que je ne m'en éloignasse que d'un étage.

J'avais entrepris, ici, de résumer brièvement les étapes, si diverses et si riches de la carrière scientifique d'André Lwoff. Je suis tenté pourtant de m'arrêter à ce point, à ces souvenirs si proches et si chers. Qui n'a pas connu le grenier à la plus belle époque mesurerait mal la richesse et la générosité de la personnalité d'André, qu'il cache soigneusement sous sa réserve un peu narquoise. Le meilleur témoignage en était cette ambiance et ce milieu qui s'étaient constitués spontanément autour de lui, et où rayonnaient la chaleur, la vie, l'imagination, l'ouverture d'esprit, le "gai savoir".

Les derniers avatars scientifiques d'André sont d'ailleurs si bien connus de la génération actuelle qu'il me paraît inutile de m'y étendre. S'attaquant, à partir de 1955 aux virus des animaux, André devait bien vite imprimer dans cette discipline sa marque personnelle et devenir l'un des maîtres incontestés de la virologie. Ce n'est après tout que sa quatrième carrière après la protistologie, la biochimie cellulaire et la Biologie Moléculaire "classique". Il y en a une cinquième: la Direction à la fois hautaine, désinvolte et familière d'un grand Institut du Cancer. Et une sixième: la Peinture. Il y révèle les mêmes dons qui ont fait de lui un homme de science exceptionnel: l'élégance, la sensibilité, la précision, le raffinement, la liberté; le style, en un mot.

Bibliographie résumée
de l'oeuvre d'André Lwoff

(1923) "Sur la nutrition des infusoires" (*C. R. Acad. Sci.*)
La première culture "pure" d'une Cilié.

(1929) "Les infraciliatures et la continuité des systèmes ciliaires récessifs," avec E. Chatton (*C. R. Acad. Sci.*)
La conception de la continuité génétique des systèmes ciliaires exprimée pour la première fois.

(1932) "Recherches biochimiques sur la nutrition des Protozoaires" (*Monographie de l'Institut Pasteur*)
La conception de l'évolution physiologique et des pertes de fonction est exprimée pour la première fois à la lumière des résultats obtenus sur la nutrition des Protozoaires libres et parasites.

(1934) "Les Ciliés apostomes," avec E. Chatton (*Arch. Zool. exp. et gén.*)
Les Ciliés parasites à deux hôtes évoluant chez les Crustacés et les Coelentérés ont été découverts. Ils présentent de remarquables métamorphoses ciliaires. La loi de la desmodexie est proposée. Continuité génétique des cinétosomes. Morphogénèse des trichocystes à partir des cinétosomes.

(1934) "Die Bedeutung des Blutfarbstoffes für die parasitischen Flagellaten" (*Zentr. Bakt. Parasitenk.*)
L'hématine, facteur de croissance, n'est pas un catalyseur. C'est une substance spécifique qui entre dans la constitution de systèmes catalytiques et agit quantitativement. La spécificité et le rôle physiologique d'un facteur de croissance sont établis pour la première fois.

(1936–1937) "Sur la nature du facteur V. Studies on codehydrogenases, I & II," avec M. Lwoff (*Proc. Roy. Soc. London*)

Le mystérieux "facteur V" est identifié aux codéhydrogénases. Son rôle physiologique pour *Hemophilus* est défini.

(1944) "L'évolution physiologique. Étude des pertes de fonction chez les microorganismes" (Hermann éd., Paris)

L'ensemble des données relatives aux besoins trophiques et au pouvoir de synthèse des Protistes considéré en fonction d'une conception de l'évolution physiologique.

(1942–1949) "Les Ciliés thigmotriches," avec E. Chatton (*Arch. Zool. exp. et gén.*)

Série évolutive de Ciliés commensaux ou parasites des Mollusques. Étude approfondie de la morphologie et de la morphogénèse.

(1947–1950) "Problems of morphogenesis in Ciliates." Dunham lectures (Harvard Medical School) (John Wiley & Sons)

La morphogénèse des Ciliés considérée dans son ensemble. Les cinétosomes ne jouent pas un rôle direct dans la morphogénèse mais sont gouvernés par les propriétés du cortex elles-mêmes sous la dépendance de la "latitude", de la "longitude" et des phases du cycle.

(1949–1950) "Lysogénie" (notes diverses) avex Antoinette Gutman, Louis Siminovitch, Niels Kjelgaard

La lysogénie établie sur une base solide. Elle est définie pour la première fois. La notion de prophage est établie, l'induction du développement du prophage, découverte.

(1953) "Lysogeny" (*Bact. Rev.*)

Développement historique de la notion de lysogénie. État actuel des connaissances. Hypothèse proposée selon laquelle les agents cancérogènes exercent leur action en induisant le développement d'un provirus.

(1955) "Control and interrelations of metabolic and viral diseases of bacteria," Harvey Lecture

Entre les maladies proprement métaboliques et les maladies proprement virales existent des processus ou états pathologiques intermédiaires.

(1957) "The concept of virus," 3rd Marjory Stephenson Lecture, London (*J. gen. Microb.*)

Une définition des virus avait été proposée dans la revue "Lysogeny." Ici, le concept de virus est discuté dans son ensemble. Existence d'une catégorie de particules infectieuses définies par l'incapacité de se diviser, la présence d'un seul acide nucléique et la reproduction à partir du seul matériel génétique. Les "petites bactéries" perdent leur droit à l'appellation de virus.

(1959) "Factors influencing the evolution of viral diseases at the cellular level and in the organism." Squibb Centennial Lectures (*Bact. Rev.*)
L'importance des facteurs non spécifiques de la résistance à l'infection virale est pour la première fois présentée dans son ensemble.

(1960) "Biological order," Karl Taylor Compton Lectures (M.I.T. Press)
Discussion des données et concepts de la biologie moderne à l'usage des jeunes physiciens et chimistes.

(1962) "Un système des virus," avec R. W. Horne et P. Tournier (*C. R. Acad. Sci.*)
Premier essai de mise en ordre d'un monde en proie au plus grand désordre.

(1965) "Specific effectors of viral development," David Keilin Memorial Lecture (*J. gen. Microb.*)
Les effecteurs du développement viral à la lumière de nouvelles expériences.

(1966) "Les effecteurs de l'infection virale primaire" (IXème Congrès International de Microbiologie, Moscou)
Les relations entre la résistance au développement viral à haute température et la virulence. Hypothèse concernant le rôle de la fièvre dans la lutte contre l'infection virale primaire.

(1969) "Death and transfiguration of a problem," Office of Naval Research Lecture (*Bact. Rev.*)
Le problème du mécanisme d'action des températures supraoptimales. Hypothèse selon laquelle ces températures agissent en provoquant la libération des enzymes des lysosomes.

André Lwoff: an appreciation

PAUL FILDES

I had not expected that it would ever fall to my lot to praise the scientific career of André Lwoff. For many years I have never read a word that he or his colleagues have published for the simple reason that their writings are beyond my grasp, and what I do know, better than some people, are my limitations.

However, it has been put to me that science is moving so rapidly that it is quite likely to blunder down the wrong alley, and in such a case I could act as an "evocation from the past." I could, for instance, deplore a tendency which is developing to "leak" discoveries to the public press before they have been established, but as a matter of fact I have not noticed this code of conduct in the sphere of cell biology. In this vast field I can only protest that the number of exponents is now so great and so diffused that one cannot attend enough dinners to get to know all of them.

One need not even be apprehensive lest André's switch to the fine arts should become detrimental to his science. Standing on the sidelines, I should say that artistry in the manipulation of a cell's insides has always been a strong feature of his work, along with a rapid acceptance of the obvious. I once, in conversation with him, expressed a view that the suggestion that a phage injected its contents into a bacterium like a syringe was too materialistic to believe. With great courtesy he asked, "How else can you explain the observed facts?" At that time I had not observed them, but I recall that I immediately accepted the syringe story. As a teacher I should think that he had a great influence in

English-speaking countries through his graceful command of the language.

I first met André Lwoff about 1934 at the Middlesex Hospital, London. I had been for many years at the London Hospital as a demonstrator of bacteriology, with, of course, no knowledge of biochemistry. As this post might be described as honorary, I was able to do quite a lot of research in my "spare time." Eventually, with funds derived from a far-seeing business, the Medical Research Council agreed to set up a "Unit for Bacterial Chemistry," which consisted of B. C. J. G. Knight, who had been with me at the London Hospital, and myself, who might be looked upon as a pupil of Knight's in biochemistry, he being a pupil of mine in bacteriology. I give these personal details because I went to the Middlesex in 1934 only to open the Unit. In 1940 the Unit went into suspended animation, and to all intents my scientific life was finished.

Thus my personal relations with André Lwoff, when I was a worker and not an evocation from the past, covered only a matter of 5 years. My attitude toward him at that time was one of admiration because he and Marguerite, who were not even bacteriologists, had discovered "growth factors" in much the same way as we had, and I felt that he (and she) had forced us to conclude that the foundations we were working on were not so insecure as some people thought.

However, when we met again after World War II, growth factor work was a thing of the past, and André had gone ahead into rarefied fields of which I knew little. Although he could, as I have said, set me right on my amateur games with phage, our relations became less centered on research and more personal. As time went on, our contacts became less frequent than we both, I think, would have liked. He went so far as to admit that he liked England and thought that I was "a typical representative of Great Britain." A footnote hoped that I would take this "for what is meant: a great compliment." I took it as he hoped, but I have a slight suspicion that he had in mind an incident in which I, as editor or policeman, had to request him to delete a quotation from Lewis Carroll inserted into a serious scientific paper. He complied with grace and so much wit that I felt like publishing the unaltered paper and the correspondence.

Not long ago the use of the English language by a French scientist in an English-speaking country was frowned upon. It is to me a striking fact that a display of nationalism has been absent in the initiation of this present project in honour of André Lwoff. The originators of this scheme, representing several nationalities and widely separated in space, have not been brought together by a common interest in science, or, I believe, solely by a common admiration of the work of André Lwoff. There is here another factor, a personal one.

On the origins of "growth factors"

B. C. J. G. KNIGHT

My contribution to this volume should be looked upon as an addendum to the paper by Sir Paul Fildes. I shall try to describe in more detail our relation with André and Marguerite Lwoff in the field of "microbial growth factors" as it was in the 1930's. I think Sir Paul and I know as much as anyone about the "microbial nutrition" phase of the work of André and Marguerite.

Fildes and I had wished to study redox potentials in connection with the germination of clostridial spores and the growth of obligately anaerobic bacteria of the genus *Clostridium*. This led us to study the growth of these organisms in chemically definable culture media and to attempt to characterize the "growth factors" needed as essential nutrients by these organisms. We transferred our attention to another pathogenic bacterium, *Staphylococcus aureus*. During this same period J. H. Mueller had been studying the nutritional requirements of still another pathogen: *Corynebacterium diphtheriae*.

To help us I collected all I could about the growth of bacteria in chemically defined media, drew general conclusions, and made some generalizations. I remember how stunned and delighted I was when, after working for some years with Sir Paul at the London Hospital and putting this material into the form of a monograph for the British Medical Research Council, I found André Lwoff's now classic publication: *Recherches biochimiques sur la nutrition des protozoaires* (1932). Bacteria and protozoa—similar physiologies?—was not then a current concept. So our ideas about microbial nutrition came together. I was able to send André proofs of my monograph after having incorporated in it references to *his* monograph, and to take his advice before mine was finally published (1936). Quite independently, we had evolved similar ideas about microbial nutrition and its reflection of microbial physiology, which shows that scientific advances can be independent because scientific publications give the same data to different individuals who may draw similar conclusions. Published scientific work is international and begets international knowledge; hence the stupidity of "secret" work about, for example, atomic bombs or chemical and biological warfare. Spies are trivial.

When I found that vitamin B1 (thiamine) and its thiazole and pyrimidine moieties together were growth factors for *Staphylococcus aureus,* I sent samples to André and Marguerite, who were working in the same field with protozoa. These synthetic compounds and analogues thereof had generously been given to us by A. R. Todd (Lord Todd) and F. Bergel, who were then working on the chemistry of vitamin B1 at the Lister Institute, Chelsea, London. Scientific workers did not then hesitate to communicate with fellow workers about their latest results. Since then the rat race has made personal scientific communication more cagey. This is a pity; those were the days!

At the same time that Fildes and I had gone to the Middlesex Hospital, thanks in part to my colleague G. M. Richardson I had identified nicotinic acid (and its amide) as another growth factor for *Staphylococcus aureus.* This discovery, coupled with the finding of André and Marguerite (1937) that cozymase 1 or 2 was the V factor required by various species of *Haemophilus,* and those of J. H. Mueller (1931 *et seq.*) about *C. diphtheriae,* stimulated more general biological interest in bacterial and protozoal "growth factors."

So was propounded the conception that certain vitamins, either synthesized by a given organism or acquired from its environment, were essential for the metabolism of a wide range of organisms. A unity of metabolic biochemistry at the cellular level was indicated. Then came the tide of war, sweeping over British science; we had other things to think about. Fildes's group was dispersed, and all that it had initiated was used by others.

By the time the war was over, science had marched on, especially in the field of "microbial nutrition" in the United States. "Microbial growth factors" had become cellular metabolism—cellular physiology and genetics.

Those are the bare bones of my connection with André and Marguerite Lwoff as scientists. May I be allowed to add some more personal remarks? During the period preceding the outbreak of World War II, I kept up a correspondence with André and still have many letters in his beautifully clear handwriting—logical, witty, human.

Around the Lister Institute, Elstree, a few wandering German planes dropped bombs, and later over the Wellcome Laboratories at Beckenham. We were on the route of V-1 fly-bombs, and Werner von Braun's V-2's. All through that time I often thought of how much worse it was for André and Marguerite. After the war my wife and I, and our children, came to know the Lwoffs very well. They were superb hosts, and Paris would not have been Paris without them.

I also well remember their visit to us when André came in 1967 to deliver the Third Marjorie Stephenson Memorial Lecture. He browsed among my books and found W. B. Yeats's *Collected Poems*. From this André chose the epigraph for his lecture:

> Things thought too long can be no longer thought
> For beauty dies of beauty, worth of worth
> And ancient lineaments are blotted out.
> *The Gyres; Last Poems* 1936–1939

In my Lwoff file is one of the most treasured possessions of my friendship with him—a splendid sketch of Rebecca Lancefield, drawn when she was delivering her Griffith Memorial Lecture to the Society of General Microbiology in 1968. I am sure that this sketch will go into some archive. It illustrates in some measure what a man of many parts is André Lwoff—artist, connoisseur, Frenchman—as well as a great original scientist.

REFERENCES

Knight, B. C. J. G. (1936). *Bacterial Nutrition: material for a comparative physiology of bacteria,* Med. Res. Council (Brit.) Spec. Rep. Series No. 210, London, His Majesty's Stationery Office, 1945; *Vitamins and Hormones,* 3, 105 (1945).
Lwoff, A. (1932). *Recherches biochimiques sur la nutrition des protozoaires,* Masson et Cie, Paris.
Lwoff, A. J., and M. Lwoff (1937). *Proc. Roy. Soc. (London),* Ser. B, 122, 352–360.
Mueller, J. H. (1940). *Bact. Rev.,* 4, 97.

Des ciliés et des hommes

(*Roscoff et Banyuls, 1923–1932*)

∽

J. MILLOT

D'autres exposeront mieux que moi, dans leur riche maturité, l'oeuvre et les découvertes de celui que nous célébrons dans ce livre. Je voudrais ici mettre l'accent sur certains aspects de sa vie scientifique débutante que j'ai été à même de bien connaître.

A.L. et moi, nous nous sommes liés à l'âge le plus favorable, celui où, sortie des brumes de l'enfance et ayant franchi les écueils majeurs de l'adolescence, la personnalité, achevant de se dégager, s'affirme dans toute sa fraîcheur où animus et anima s'essayent à vivre, mettant spontanément à l'épreuve leurs jeunes forces encore maladroites, que les blessures de l'existence et l'usure de l'expérience n'ont encore ni affaiblies, ni découragées. L'amitié hors du commun qui en est résultée n'a pas à être définie: du moins me permet-elle d'évoquer, non sans quelque émotion, les milieux scientifiques initiaux où s'est développée l'originalité d'A.L., d'esquisser les conditions de ses premières recherches dans les laboratoires de Roscoff et de Banyuls, rendant, chemin faisant, hommage à quelques uns des personnages qui, à divers titres, en furent les guides ou les témoins.

ROSCOFF

Mes souvenirs remontent à mon premier séjour au Laboratoire maritime de Roscoff, en 1923. Cet établissement fondé un demi-siècle auparavant, sous une forme modeste, par Lacaze-Duthiers, qui souhaitait contribuer ainsi au relèvement de la France après le désastre de 1870–1871, avait acquis une grande réputation parmi ceux qui s'intéressaient à la biologie marine: celui qui en assurait la direction, sous l'autorité suprême de Charles Pérez, était Marcel Prenant, fils d'Auguste Prenant chez qui j'apprenais l'histologie à la Faculté de Médecine de Paris; par lui, le nom et les vertus de Roscoff m'étaient devenus familiers et, zoologiste de coeur, je voulus aller passer mes vacances d'été dans ce temple de la science marine. J'y trouvai A.L. Quoique plus jeune, il m'avait largement devancé, y ayant fait un premier stage dès 1920, tout juste âgé de 18 ans, et y étant retourné chacune des années suivantes.

Le cadre était celui d'un petit port breton caractéristique, avec ses maisons et son église de granit, judicieusement choisi comme siège d'un grand Laboratoire maritime du fait de l'amplitude des marées, alliée à la variété et à la richesse biologique des faciès littoraux, où coexistent à profusion rochers, sable, herbiers et bancs de vase. Population calme de pêcheurs, débordée pendant la belle saison par la venue des travailleurs du Laboratoire et de ce que ceux-ci qualifiaient dédaigneusement de "goitreux", c'est-à-dire l'humanité inférieure de touristes et de vacanciers dont les hôtels et les quelques villas disponibles se partageaient le modeste afflux. Les environs, où dominaient champs de pommes de terre, d'artichauts ou de choux-fleurs—importants produits d'exportation vers l'Angleterre—étaient esthétiquement fort ingrats, et l'horizon marin, barré par l'île de Batz, n'offrait lui-même à la contemplation que des lignes peu harmonieuses. Mais dans ce paradis de la zoologie marine, toutes les conditions favorables au travail libre et sain se trouvaient réunies.

Le Laboratoire, considérablement agrandi au cours des dernières décennies, est certes devenu bien différent de celui que nous connûmes il y a plus de 40 ans. Il nous paraissait déjà alors fort imposant avec ses deux ailes perpendiculaires, l'une en façade sur la mer, l'autre bordant la place de l'Église, groupant, avec les stalles de travail, les magasins, la bibliothèque, le vaste aquarium à eau de mer courante, formant galerie vitrée, complété par une terrasse et un vivier: un peu en dehors enfin, d'un côté, la maison du Directeur, de l'autre, un petit bâtiment servant de salle de récréation pour les travailleurs, où avaient lieu "initiations", soirées costumées et mille autres fantaisies.

En dehors des conférences de perfectionnement destinées aux étudiants, le travail s'organisait pour les chercheurs sans règle propre, suivant l'horaire et selon les modalités les plus convenables aux habitudes et aux sujets de chacun: tantôt prédominaient l'exploration des grèves ou des herbiers à marée basse et la détermination des échantillons recueillis, tantôt les recherches anatomiques ou microscopiques en stalles, tantôt la mise en train d'élevages ou d'expériences dans les bacs de l'aquarium.

Le Laboratoire voyait déferler l'été une bonne partie de la zoologie française et étrangère. Des maîtres chevronnés—Jean Cantacuzène, Paul Wintrebert, Georges Bohn, Emmanuel Fauré-Fremiet, Édouard Chatton, Constantin Davydoff, Fred Vlès, Paul Marais de Beauchamp, pour ne citer que quelques-uns des plus fidèles—y rencontraient de nombreux débutants à divers stades. Les uns, parmi les mieux doués, devaient disparaître prématurément et parfois tragiquement: je pense surtout ici à Michel Volkonsky, dont la perte fut si cruelle à ses amis,[1] à Maurice Azéma, tué dans un accident de montagne, à Maurice Parat, noyé dans le naufrage du *Pourquoi-Pas,* à Louis Rapkine. D'autres devaient devenir eux-mêmes des maîtres éminents, égalant sinon surclassant ceux qui les avaient formés: il suffit de rappeler parmi les compagnons proches ou distants d'André Lwoff, de 1920 à 1930, les noms de Robert Courrier, Boris Ephrussi, Maurice Fontaine, Jean Piveteau, Marcel Roubault, Georges Teissier, Étienne Wolff, René Wurmser.

Il est difficile de se représenter, à notre époque de contestation dans la violence et de désarroi universitaire, l'agrément et le profit intellectuels et humains que l'on pouvait alors retirer de ces stages en semi-communauté dans les grands laboratoires maritimes, la totale indépendance dans le travail, l'harmonieux équilibre entre la recherche et les distractions. Le brassage dans la confiance et la bonne humeur n'excluait en aucune façon le respect ou la critique entre biologistes de tous âges, étudiants ou maîtres affirmés. On ne saurait compter le nombre de vocations de zoologistes ainsi éveillées. Chatton a lui-même écrit que son premier séjour à Roscoff, en 1902, fut pour lui "la révélation d'un monde nouveau" qui décida de l'orientation de sa vie scientifique. Deux extraits de lettres d'Auguste Prenant ressuscitent, avec humour, les deux aspects,

---

[1] André Lwoff, Michel Volkonsky, *Ann. Inst. Pasteur,* **68**, 489 (1942).

complémentaires plutôt que contradictoires, de la vie du Laboratoire au temps où y travaillait A.L.:

"Le Laboratoire est un congrès permanent ou chacun veut entrenir "l'autre de ses occupations, lui montrer ses préparations. Et chacun est "un multiple de n. Il y a, en outre, Pérez, Cantacuzène, Wintrebert, "Chatton, Vlès, Davydoff pour ne citer que les huiles lourdes, sans "compter les demi-lourdes et en négligeant les légères: une demi-lourde "du nom de Witschii m'a instamment prié de compter les chromosomes "dans une prétendue mitose de maturation. Je lui ai répondu—et il a "paru surpris que je n'étais pas venu à Roscoff pour compter les chromo- "somes, ayant fait ce métier toute l'année. Mais j'ai compté tout de "même" (22 août 1924).

"J'ai fait ici deux recrues, l'une Dimitriescu, élève de Cantacuzène ". . . l'autre nommée de Rogalsky, polonais . . ." "Le Laboratoire "est en fête perpétuelle et on y sable le champagne dans un casino privé "jusqu'à une heure avancée de la nuit. Je préfère me coucher presque "avec les poules, pour pouvoir être de ceux qui, le lendemain, au réveil, "poussent un cocorico zoologique retentissant" (6 septembre 1925).

Dans la correspondance d'A.L. lui-même, quelques formules lapidaires qualifient ses journées roscovites: "Infusoires, Infusoires, avec alternance de Crustacés mayonnaise" . . . "Travail continu depuis un mois. Aucune lecture. Bains de soleil—lancement de cailloux de forme variable—combats de pommes de terre—petit jeu" . . .

Cette dernière citation incite à décrire plus en détail les distractions de Roscoff.

Dans la journée, celles-ci ne manquaient pas. Aux conversations de plein air, par petits groupes formés au hasard des sympathies, réunis soit sur une des places publiques, devant quelque hôtel en prenant le café, après le déjeuner (Fig. 1), soit dans le jardin du Laboratoire, soit sur sa terrasse, soit encore en bordure de mer, s'ajoutaient les promenades dans les environs du bourg, assez réduites il faut l'avouer et de peu d'intérêt par elles-mêmes. Exceptionnellement étaient organisées des excursions collectives en autocar, ou de courtes croisières en mer sur le bateau du Laboratoire, le "Pluteus", dont la stabilité assez incertaine exposait les plus augustes passagers aux affres démoralisantes du mal de mer.

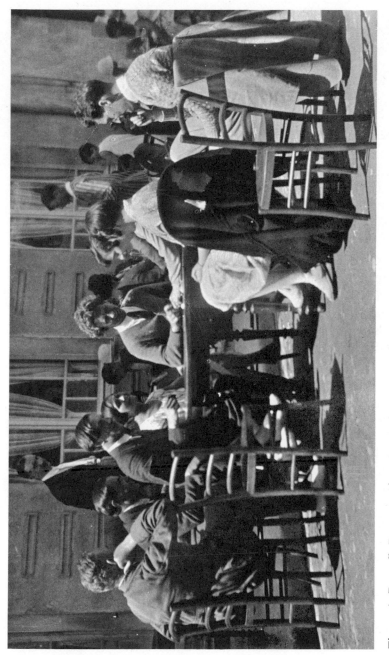

Figure 1. Roscoff. Pousse-café après le déjeuner sur une place près du laboratoire. De gauche à droite: Georges Teissier de profil, Henriette Millot de profil à demi-cachée, André Lwoff de face au milieu, Marguerite Lwoff à sa droite; à côté, Boris Ephrussi; derrière, Neukomm de profil (Cl. J. Millot, 1927).

Les bains, qu'ils soient de mer, ou de soleil les jours favorables, étaient, comme on peut le supposer, une des occasions les plus courantes de délassement.

"Faire la marée" était aussi, il va de soi, une activité classique à la fois instructive et récréative, elle contribuait à assurer le ravitaillement des travailleurs en animaux d'étude et à enrichir leur connaissance des biotopes marins; elle alliait aux plaisirs de la chasse et de la découverte de spécimens rares, les vertus d'un exercice sportif non négligeable.

Les vrais sports et les jeux restaient limités. Le tennis était roi: nous étions nombreux à le pratiquer. On y admirait la forme physique de Paul Wintrebert: quoique Professeur grisonnant, celui-ci restait, à près de 60 ans, la meilleure raquette du Laboratoire. Sur un plan plus familier, le jeu de boules avait ses adeptes: je l'ai pratiqué avec Maurice Fontaine de Marcel Roubault, mais ne me souviens pas d'y avoir affronté André Lwoff. Par contre, nous nous livrions souvent à des concours de ricochets de galets lancés tangentiellement sur l'eau: Lwoff et Volkonsky y excellaient. D'autres jours, nous nous adonnions à des batailles de pommes de terre, dont nous trouvions parfois sur les plages des amas résiduels importants. Une distraction classique, celle peut-être dont nous nous lassions le moins, était le "petit jeu". Cet exercice d'adresse se pratiquait entre deux personnes de taille et de poids proportionnés, placées face à face, les pieds solidement implantés sur le sol ne devant pas être déplacés: le but était de se faire réciproquement perdre l'équilibre en se heurtant fortement le plat des mains, sans se toucher autrement. Le plus habile l'emportait aisément sur le plus vigoureux, souvent incapable de maîtriser l'excès de ses élans: il en résultait des attitudes et des redressements fort pittoresques (Fig. 2).

Les divertissements nocturnes nécessitaient plus d'élaboration. Des soirées costumées étaient organisées de temps à autre: de peu d'apparence, certes elles devaient leur mérite à la pauvreté même des moyens utilisables et au temps fort restreint disponible pour leur préparation. Cette limitation extrême des ressources stimulait l'ingéniosité de l'improvisation dans la bonne humeur.

Les "séances d'initiation", dites aussi de "zébrification", étaient traditionnelles. Marcel Prenant les présidait le plus souvent. La victime consentante s'y trouvait soumise à une succession d'épreuves ou d'interrogatoires, à tendances humoristiques, ne témoignant pas toujours d'un

Figure 2. Roscoff. Le "petitjeu": Boris Ephrussi et André Lwoff s'efforçant de garder l'équilibre (1927).

très haut niveau intellectuel, mais auxquelles l'imprévu des résultats et la bonne volonté générale assuraient une ambiance détendue et distrayante. Aux exercices d'adresse pittoresques (ramassage d'objets avec les dents, capture, les yeux bandés, d'un jeune congre dans un baquet, etc.), succédaient des énigmes baroques ou des questionnaires burlesques. Un diplôme de fantaisie récompensait ceux qui avaient triomphé de toutes les difficultés et avaient su passer par le "trou du Zèbre".

L'humour roscovite se manifestait également dans les chansons élaborées au Laboratoire, auxquelles Charles Pérez lui-même ne dédaignait pas de collaborer. La "Complainte du Directeur" est restée célèbre.

Une autre, évoquant les activités du Laboratoire, mettait familièrement en cause André Lwoff, sur l'aire de Cadet Rousselle:

"Dédé taquine le Protiste (bis)
"Jusqu'à c'qu'il se fourr' sous un kyste (bis)
"Puis lui dit 'A l'année prochaine
" 'Je verrai si ton cycl' s'enchaîne'
"Ah! Ah! Ah! oui vraiment
"L'Cilié d'Roscoff est épatant."

La chanson du "Secret de la vie", due à Georges Teissier, était plus philosophique: un savant ayant découvert le secret de la vie, propose aux différents chercheurs du Laboratoire de le leur communiquer: tous successivement, constatant que ce secret ne concerne pas directement leurs propres recherches sur les Oursins, les Rotifères ou les Protozoaires, refusent de s'y intéresser.

André Lwoff arrivé à Roscoff en été 1920, fraichement nanti de ses deux premiers certificats de licence, y reçut l'enseignement de Fred Vlès. Revenu l'année suivante, sa licence terminée, il se fit, par ses dons intellectuels évidents et par son ardeur au travail, remarquer de Lucienne Dehorne, alors chef de Travaux au Laboratoire. Celle-ci le recommanda à Édouard Chatton, qui cherchait un jeune coéquipier pour ses recherches de Protozoologie et avait mission de recruter pour l'Institut Pasteur un Assistant de qualité. La collaboration s'instaura d'emblée avec plein succès. Avant la fin de l'année, une note en commun paraissait à l'Académie des Sciences sur une nouvelle famille d'Acinétiens de Mollusques acéphales, prélude d'innombrables découvertes. Cette heureuse rencontre décida des débuts de la vie scientifique d'André Lwoff et fit naître une des plus fécondes associations zoologiques de notre temps.

Édouard Chatton était en 1921 en pleine forme créatrice. Agé de 38 ans, Maître de Conférences à la Faculté des Sciences de Strasbourg en attendant d'y devenir Professeur titulaire, auteur d'une thèse magistrale sur les Flagellés Péridiniens, il se spécialisait dans la Protozoologie marine et était particulièrement intéressé par les formes parasites. Esprit ferme et coeur simple, quoique toujours insatisfait, animant un corps puissant, tout en muscle, ce travailleur opiniâtre et scrupuleux ne publiait que des travaux irréprochables: excellent dessinateur, il les illustrait lui-même avec talent.

L'aide d'André Lwoff, la parfaite entente qui s'établit très vite entre eux et ne cessa de s'amplifier, devenant une amitié profonde entre disciple et maître, permit à Chatton de donner toute sa mesure. Celui-ci fut, en revanche, pour André Lwoff à ses débuts un initiateur incomparable, un entraîneur remarquable et un guide scientifique sûr. Leurs journées de travail à Roscoff, où ils se retrouvèrent chaque été jusqu'en 1929, plus irrégulièrement ensuite, se prolongeaient habituellement fort tard: bien souvent, ils étaient le soir les derniers à quitter le Laboratoire, quand ils n'y revenaient pas après diner. Ainsi, à coup d'observations ingénieuses et méthodiques inlassablement poursuivies, de trouvailles s'enchaînant les unes les autres, ils élaborèrent sur les Infusoires ciliés, dans "une atmosphère exaltante d'offensive victorieuse", une oeuvre qui transforma notre connaissance de ce grand groupe et fait honneur à la zoologie française de ce siècle. Marguerite Lwoff devait bientôt y être très activement associée. Les découvertes les plus retentissantes concernent les sous-ordres des Ciliés Thigmotriches, parasites des branchies de Mollusques Lamellibranches et de Tuniciers, dérivés des Ciliés par une série de dégradations morphologiques, finalement devenus glabres et fixés sur leur hôte par un long suçoir: leur étude par Chatton et Lwoff a apporté des faits et des interprétations d'un haut intérêt pour la compréhension des êtres unicellulaires. La mise en évidence du groupe des Apostomes fut également un des évènements de la Protozoologie: les cycles biologiques complexes et variés de ces Ciliés, identifiés à partir de kystes dans des Crustacés Copépodes, n'évoluant qu'après ingestion par un Hydraire, et donc à hôte double, furent démêlés avec une ingéniosité remarquable et permirent d'étayer des considérations inédites sur l'ontogénèse du groupe. Faut-il rappeler encore la découverte des Pilisuctoridés, parasites des poils de certains Crustacés Amphipodes, sur lesquels les embryons de ces Ciliés s'empalent activement par la bouche, se nourrissant de leur secrétion?

Mais aux yeux de nos auteurs, on ne saurait trop le souligner, les études de morphologie, si attachantes et fécondes qu'elles pussent être, ne prenaient toute leur valeur qu'en tant que prétextes et moyens d'aborder et de résoudre des problèmes d'ordre général de cytologie, de développement, ou de phylogénie.

Édouard Chatton comptait parmi les personnalités dominantes du Laboratoire où son caractère réfléchi, sa solidité d'homme de l'Est, son ardeur et sa maîtrise dans le travail, ses réactions franches et directes, l'impression de force et d'énergie qui émanait de sa physionomie[2] le faisaient tenir en respectueuse considération. Faute de meilleure anecdote digne d'être rapportée à son sujet, j'évoquerai le soir où, étant dans un café avec Paul Wintrebert, de vingt ans son aîné et de silhouette plus fragile, celui-ci se prit de querelle avec un voisin qui prétendit lui "tirer les oreilles". "Vous voudrez bien commencer par me tirer les miennes" déclara aussitôt Chatton, s'avançant d'un air résolu, qui fit immédiatement battre en retraite l'imprudent.

Quant à André Lwoff, comment apparaissait-il à Roscoff dans cette faune de travailleurs de tous âges et de toutes origines? Physiquement déjà, il sortait de l'ordinaire: grand, blond, auréolé d'une abondante chevelure dorée, semi-bouclée, il pouvait évoquer, lorsqu'il inclinait la tête avec le léger sourire qui lui était familier, l'ange célèbre de la Cathédrale de Reims. Cet "angélisme", nuancé d'une apparente nonchalance, n'était pas exempt non plus d'une certaine allure romantique, accentuée par le veston de velours noir que notre héros aimait à porter à Roscoff.

Ses dons intellectuels précoces se manifestèrent surtout à Roscoff par une exceptionnelle aptitude à l'observation et à l'interprétation des plus minimes détails structuraux des Protozoaires tels que les révélaient les microscopes de l'époque, ainsi qu'à la reconstitution des cycles évolutifs des êtres unicellulaires, en attendant de s'appliquer, par une évolution progressive et toute naturelle, à l'analyse et à la compréhension synthétique des mécanismes les plus complexes de la biologie cellulaire. La nonchalance trompeuse du comportement masquait un oeil et un esprit critique toujours en éveil, accompagné d'une grande facilité et, s'il le fallait, de beaucoup d'acharnement dans le travail.

Hors du domaine scientifique, André Lwoff n'était, en littérature comme en musique, indifférent à aucune oeuvre de qualité, mais, dès sa jeunesse, sa vive sensibilité esthétique l'orientait particulièrement vers les arts plastiques, sculpture (il modela un excellent bas-relief de

---

[2] Se reporter au portrait de Chatton tracé par A.L. lui-même: *Arch. zool. exp. gén.*, Notes et Revues, **85**, 121–137 (1948).

Chatton) et surtout peinture, destinée à devenir le plaisir dominant et la nécessaire évasion de sa maturité, l'antidote de cette "science inhumaine" qui devait faire sa célébrité, mais s'avérait incapable de satisfaire ses exigences affectives les plus profondes. Cet amour de la beauté, ce besoin d'absolu, s'affirmant presque désespérément avec l'âge par l'expression picturale, devaient, suivant sa propre expression, le "submerger" à l'heure même de ses plus grands succès scientifiques et

Figure 3. Roscoff. Dans une salle de travail. Debout: Boris Ephrussi et E. Fauré-Fremiet; assis par terre: Henriette Millot et André Lwoff (1927).

lui faire regarder la création artistique comme la forme de résistance la plus valable aux forces de destruction et de mort.

Moralement—s'il n'existe pas de meilleur terme pour désigner l'immense part d'un être humain qui n'est ni directement physique, ni du domaine propre de l'intellect, cette "anima" qui commande le caractère, les réactions sensibles et fait de chacun de nous l'essentiel et le plus authentique de ce qu'il est—c'est le mot distinction qui, dès les premiers temps où je l'ai connu, pouvait le mieux caractériser André Lwoff.

Répulsion innée devant toute bassesse, mesquinerie ou vulgarité, simplicité réelle de réactions aristocratiques naturelles, parfois interprétées trop légèrement comme trahissant un tempérament orgueilleux, goût intransigeant du travail bien fait, coexistaient en lui avec un désintéressement peu commun envers les problèmes de carrière aussi bien que d'argent: il ne s'est jamais caché de ne pouvoir estimer ceux de ses pairs dont des considérations de cet ordre commandaient, de façon trop évidente, les activités. Ce peu de souci des avantages matériels était allié à une générosité foncière, manifestée aussi bien sur le plan de la vie personnelle que dans le domaine social. Après plus de quarante années, je n'ai pas oublié son émotion indignée à Roscoff lors du procès des anarchistes italiens Sacco et Vanzetti, et sa solidarité avec toutes les victimes d'oppressions injustes ne s'est jamais démentie.

Faut-il insister sur son besoin fondamental d'indépendance? A.L. a toujours su s'assurer, vis-à-vis du monde comme de lui-même, cette liberté dont bénéficient ceux qui s'imposent, spontanément ou par raison, les disciplines nécessaires à toute vie efficace. Son refus de souscrire aux conformismes de principe, aux servitudes sociales abusives, à des démarches ou à des usages professionnels courants mais à ses yeux insuffisamment justifiés, s'est affirmé dans l'âge mûr, mais il s'annonçait déjà dans la jeunesse: j'admirais d'autant plus le courage et la pureté de ce comportement que je n'avais pas toujours moi-même la force d'âme de me dégager de bien des contraintes individuelles ou collectives, que cependant je supportais mal.

Dans tous les domaines, le courage a d'ailleurs été une des vertus d'A.L. qu'il l'ait manifesté au cours des premières années de sa vie scientifique par son élégance à s'accommoder de conditions d'existence fort modestes, plus tard par son rôle dans la Résistance, surtout peut-être, tout au long de sa route, par son refus de tout calcul ou de toute compromission, le conduisant parfois à exprimer sans ménagements, sur les gens ou les choses, les critiques qu'il croyait justifiées, en se souciant peu des mécontentements qu'il avait chance de provoquer. C'est que sa douceur de caractère, son abord facile et son commerce plein d'agrément, n'excluaient pas, le cas échéant, une volonté inébranlable.

L'essai de portrait ci-dessus exigerait bien des touches complémentaires, et que fussent signalées comme définissant aussi A.L., une vue souvent ironique des êtres et des choses alliée à beaucoup de tolérance—un humour parfois caustique, mais toujours indemne de méchanceté—une sagesse de l'âme, s'acceptant sans débats de conscience, ne dissipant pas temps et forces en vains scrupules ou en stériles regrets d'actes passés—une appréciation positive des valeurs humaines, associée à une absence d'inclinations religieuses le disposant, à Roscoff, à juger "déprimantes" les manifestations extérieures du christianisme breton—enfin dans l'amitié et les rapports humains personnels, une délicatesse, une sensibilité et une discrétion, sur l'évocation desquelles on me permettra, sans y insister d'avantage, de conclure ces quelques pages.

BANYULS-SUR-MER

Quelle que fût la féconde activité de l'équipe Chatton-Lwoff à Roscoff et l'importance des découvertes qu'elle sut y réaliser en Protozoologie, son oeuvre, s'est brillamment poursuivie et notablement étendue au Laboratoire Arago de Banyuls, rival catalan de l'Institut breton et, comme lui, une des plus heureuses réalisations de l'Enseignement supérieur français à la fin du XIXᵉ siècle.

D'importance comparable, voués aux mêmes tâches, les deux Etablissements se suppléaient, en s'opposant à maints égards, chacun ayant sa personnalité propre. Tout, ou presque, contrastait entre eux: non seulement le climat chaud et sec, parfois brûlant du Roussillon, bien différent de celui de la douce et humide Bretagne, transformait le calendrier de travail, les "saisons" prédominantes de Banyuls, printemps et automne, ne concurrençaient aucunement la grande "campagne" d'été à Roscoff—mais à Banyuls, l'arrière-pays fort accidenté

Figure 4. Banyuls. Leçon de pipettes dans la stalle de l'Assistante d'alors du Laboratoire, Odette Tuzet, depuis Professeur à la Faculté de Montpellier (Cl. J. Millot, 1931).

et de beaucoup de caractère, habillé de vignes, d'orangers, d'oliviers, de pins, de chênes verts ou de chênes-lièges, l'éclat de la lumière, les senteurs des garrigues distillant leurs essences au soleil, le pittoresque des vendanges à l'espagnole, l'opposition des tuiles méditerranéennes et des ardoises bretonnes, du tempérament catalan confronté à celui du Finistère, jusqu'au type physique des femmes immortalisé dans la pierre par le sculpteur Maillol, grand homme du lieu, tout contribuait à créer au Laboratoire Arago un milieu plein d'attraits, auprès duquel Roscoff pouvait paraître triste et morne à certains.

De telles divergences, déjà si marquées dans le domaine des conditions naturelles, étaient encore accentuées sur le plan psychologique par plusieurs facteurs. La situation géographique de Banyuls, la grande distance séparant Paris de la Catalogne et la lenteur du chemin de fer de l'époque, commandaient le recrutement des travailleurs: ceux-ci, provenant en majorité du Sud et de l'Est de la France, voire de Suisse, comprenaient beaucoup plus de provinciaux que de parisiens et l'atmosphère humaine s'en ressentait fort. Le Directeur, Octave Duboscq, lui-même d'origine et de formation provinciales, profondément attaché à ses amis comme à ses habitudes, avait marqué la vie du Laboratoire par l'austérité et la régularité rigide de sa propre existence et le sérieux de son caractère. A aucun niveau, l'ironie normalienne, le goût du canular, toujours sous-jacents à Roscoff, ne se faisaient sentir à Banyuls. Si la beauté du paysage faisait rêver les travailleurs au point de détourner certains d'entre eux de leurs tâches, les distractions se bornaient aux bains de mer dans les charmantes petites criques côtières ou dans l'eau tiède de la baie qui "portait" admirablement, et à des promenades variées et admirables: celles-ci demandaient à pied un peu de souffle et de jarret, même pour la seule forêt de la Massane, mais on pouvait en automobile gagner, en peu de temps, d'aussi beaux sites que St Martin du Canigou ou explorer largement la Catalogne espagnole. Le soir, certains des hôtes du Laboratoire se réunissaient pour entendre des disques dans l'appartement de l'Assistant d'alors, André Migot, sportif séduisant, passionné de montagne, dont rien ne faisait alors prévoir les conversions successives à la vie monastique cistercienne, puis au bouddhisme.

Chatton était venu travailler dès 1905 au Laboratoire Arago, dont il devait être nommé Directeur en 1937, succédant à Duboscq. Il avait mis en évidence à son premier séjour un Péridinien polyblastique fort singulier, chez des Copépodes pélagiques. Cette découverte décida de toute l'orientation de ses recherches et fut à l'origine de l'amour violent et presque exclusif qu'il devait, par la suite, manifester pour les Protozoaires. C'est à Banyuls, de 1907 à 1913, que fut en majeure partie élaborée sa monumentale thèse de sciences.

André Lwoff fut attiré par son maître au Laboratoire Arago à partir de 1923. Sur cette faune méditerranéenne originale et si différente de celle de l'Atlantique, ils purent l'un et l'autre exercer au mieux leur remarquable aptitude à identifier les innombrables organismes monocellulaires des eaux marines, à en analyser les structures les plus

Figure 5. Paris. André Lwoff mobilisé à l'hôpital Villemin, tout sombre d'être
sevré d'Infusoires (Février 1928).

significatives, à en pressentir intuitivement, puis à en établir les cycles évolutifs, souvent inattendus. Ainsi accumulèrent-ils les découvertes.

Une très large part des recherches sur les Apostomes fut, en particulier, accomplie à Banyuls. Lorsqu'en 1928 André Lwoff, retenu par ses obligations militaires (Fig. 5), ne put se rendre, comme à l'accoutumée, au Laboratoire Arago, Chatton en fut véritablement consterné; il réquisitionna en quelque sorte Marguerite Lwoff pour compenser cette malencontreuse absence afin que la monographie des Apostomes, en si bonne voie, ne risquât pas d'être trop retardée.

Ainsi, dès avant cette époque, le rôle d'André Lwoff dépassait de beaucoup celui d'un brillant second. Vite hissé de la condition d'élève à celle d'associé à part entière, il était devenu, dans la célèbre équipe, un élément moteur essentiel, en attendant de développer glorieusement sa jeune maîtrise et de se tourner vers d'autres horizons a l'Institut Pasteur.

Aux croisements de nos chemins

❧──❦──❧

EMMANUEL FAURÉ-FREMIET

I. OÙ DE VIEUX ET CHERS SOUVENIRS SONT ÉVOQUÉS

Je ne saurais dire comment j'ai connu Édouard Chatton, lorsque, voici plus de soixante années, nous abordions l'un et l'autre, en des lieux différents, l'apprentissage du laboratoire. Sans doute nos premières rencontres doivent-elles se situer d'une part sous les combles vétustes de la chapelle des Cordeliers, rue de l'Ecole de Médecine, où la Société de Biologie tenait Séance chaque samedi, et de l'autre dans la triste salle de la rue Serpente où la Société Zoologique de France se réunissait deux fois par mois. Nos premières publications, sensiblement contemporaines, sont datées des lointaines années 1904 et 1905.

Sitôt promu licencié ès Sciences naturelles, Édouard Chatton, mon aîné de trois mois, avait commencé, en 1905, à la Sorbonne, dans le laboratoire du Professeur Pruvot, l'étude de quelques Protozoaires. Ses studieuses vacances d'été le conduisirent au Laboratoire de Banyuls, et c'est là qu'il découvrit, chez des Copépodes marins, un groupe nouveau de Dinoflagellates curieusement adaptés à la vie parasitaire. Servi par un sens aiguisé de l'observation et par une opiniâtreté à toute épreuve, Chatton parvint à reconstituer, à travers diverses transformations imprévues, le cycle évolutif de ces organismes aux aspects parfois déconcertants. C'était un premier et remarquable succès.

Figure 1. Édouard Chatton: expression puissante et volontaire, qui n'exclue pas nécessairement l'inquiétude et le doute.

De mon côté,[1] c'est en microscopiste amateur, épris des choses de la vie, que, tout en me préparant à une autre carrière, j'avais consacré maints loisirs de mes années de jeunesse à l'observation attentive des Infusoires Ciliés. Les "Leçons sur la cellule" de L. F. Henneguy étaient l'un de mes livres de chevet, et lorsque ce Maître voulut bien m'accueillir à son laboratoire du Collège de France, je pus contrôler et compléter mes observations à l'aide de conseils éclairés et de techniques cytologiques adequates. La chance m'avait servi en me faisant constater la présence constante, dans le cytoplasme, de particules se multipliant par bipartition, et que je devais bientôt identifier au mitochondries de Benda; elle m'avait comblé en éclairant la morphologie comparée des Ciliés Peritrichida par la découverte de leur scopula, organite fixateur, constitué par des cils modifiés; ce qui fut confirmé, 50 années plus tard, par la microscopie électronique.

Parfois, avec notre ami Paul de Beauchamp, jeune maître des Rotifères, nous avons, Chatton et moi, exploré quelques mares dans la paisible vallée de Chevreuse, ou sur le plateau de Villejust, aux abords duquel nous déposait l'"Arpajonnais", pittoresque et crasseux petit train sur route. Après quelque longue et fructueuse randonnée, attendant le train du retour, nous nous attablions chez le bistrot du coin; c'est alors que, avec calme et sérieux, de Beauchamp nous disait "je ne bois pas" et partait à l'aventure. Peu avant l'heure du départ nous voyions reparaître, avec un large sourire, cet incomparable naturaliste; il avait su, en quelques minutes, trouver *le* ruisseau, soulever *la* pierre, et capturer *la* bestiole rare, provoquant ainsi notre cordiale et chaleureuse admiration.

C'est au cours de ce temps que les connaissances zoologiques de Chatton et la maîtrise avec laquelle il avait su résoudre de difficiles problèmes protistologiques, l'appelèrent à l'Institut Pasteur, en 1907, dans le service de Protistologie et Microbiologie coloniale fondé par Laveran et dirigé par Félix Mesnil; et quelques années plus tard, c'est

---

[1] Ecrivant ces lignes au cours de sa 86ème année leur auteur croit devoir signaler au lecteur que les progrès de l'âge entraînent fréquemment une facheuse tendance à (1o) trop parler de soi lorsque l'on se propose de parler des autres; (2o) oublier ces paroles d'Elihu, fils de Barachel le Bouzite: "Ce n'est pas l'âge qui procure la sagesse, ce n'est pas la vieillesse qui rend capable de juger" (Job 32:9).

auprès de Charles Nicolle qu'il devait travailler à l'Institut Pasteur de Tunis. Dès lors nos deux chemins s'écartèrent et, sans affaiblir notre amitié, nos rencontres se firent beaucoup plus rares.

Tandis qu'Édouard Chatton abordait avec un plein succès les nombreux problèmes posés par les Protistes pathogènes, Trypanosomes, Toxoplasmes, Amibes, etc, comme par leur mode de pénétration, et leur cycle évolutif, je faisais de sérieuses infidélités aux Infusoires en abordant avec enthousiasme quelques problèmes de cytochimie. Ce fut, en particulier, avec André Mayer et Georges Schaeffer, la détection des lipides mitochondriaux, puis, utilisant l'Ascaris du Cheval, l'isolement et la purification des constituants de certains corps figurés présents dans l'oeuf et dans le spermatozoïde, et l'étude de leurs transformations au cours de la fécondation.

Cependant, les vacances nous ramenaient chaque été, l'un et l'autre, vers les Protistes marins; Chatton à Banyuls, à Villefranche, à Roscoff et moi même au Croisic, auprès de mon maître Henneguy et plus tard, à Concarneau.

Vint la dramatique épreuve des années 1914–1918; puis l'après-guerre et la réorganisation du travail de recherche. Édouard Chatton fut nommé d'abord Maître de Conférence à l'Université de Strasbourg, où, en 1922, il succédait au Professeur Bataillon, à la chaîre de Biologie générale; puis, en 1932, il passait à la Faculté des Sciences de Montpellier, ce qui lui permit de réorganiser et d'utiliser largement le laboratoire maritime de Sète.

II. OÙ IL SERA QUESTION DE TECHNIQUE

Les Protistes aussi petits et aussi peu sympathiques que les Trypanosomes s'accommodent parfaitement d'un examen post-mortem utilisant les techniques ordinaires des colorations cytologiques. Par contre, l'étonnante complexité structurale réalisée à l'échelle unicellulaire par les Infusoires Ciliés, et la réelle beauté de ces minuscules organismes, ne sont pleinement sensibles qu'à l'état de vie active. Souvent méconnaissables après l'action des réactifs fixateurs, ces admirables bestioles ne tolèrent les multiples raffinements de la technique que pour la mise en évidence des structures cytoplasmiques et nucléaires dans leurs ultimes détails.

On remarquera d'autre part que la forme d'un Cilié, avec toutes les particularités caractéristiques de l'espèce, est intimement liée, et comme dominée, par la distribution spatiale de ses organites vibratiles: cils, soies, cirres ou membranelles. Les figures publiées au siècle dernier par un Stein, un Lieberkühn, un Maupas, provoquent l'admiration devant la virtuosité de ces anciens Maîtres poursuivant l'examen d'un Cilié vivant et mobile sous tel microscope monoculaire dont l'optique était bonne, mais dont aucun débutant n'accepterait l'usage aujourd'hui. Un peu plus tard, la pratique des coupes à la paraffine et des colorations à l'hématoxyline ferrique, permirent de mettre en évidence, chez les Infusoires, les corpuscles basaux des cils vibratiles et leur disposition fréquente en rangées longitudinales; ces résultats complétaient ou précisaient beaucoup de descriptions précédentes obtenues avec moins de sureté. Il n'est pas inutile de rappeler ici le beau travail publié en 1903 par H. N. Maier sous le titre "Über den feineren Bau der Wimperapparate der Infusorien."

Cependant, la distribution des cils à la surface du corps de l'Infusoire restait, dans bien des cas, insuffisamment définie, et les mécanismes assurant la réorganisation et la bipartition restaient incomplètement connus.

C'est en 1926 que Bruno Klein utilisa la technique d'imprégnation au nitrate d'argent, par laquelle Ranvier mettait en évidence les limites cellulaires, pour examiner la structure corticale des Ciliés. Cette fois les corpuscles basaux et leurs assemblages étaient mis en évidence avec une précision inégalée jusqu'alors; mais la cruelle nécessité de dessécher au préalable les Ciliés sur le porte objet altérait profondément leur forme générale, ce qui nuisait lourdement à l'interprétation des images argentiques.

III. OÙ L'ON VOIT APPARAÎTRE UN GRAND ET TIMIDE(?) JEUNE HOMME

Cela se passait en l'été 1920 et commençait dans la petite presqu'île qui sépare la baie de Morlaix de l'estuaire de la Penzé; plus exactement

Figure 2. André Lwoff: Expression malicieuse d'euphorique ironie, qui n'exclue pas nécessairement de profondes sensibilités affectives.

à Carantec, où le Dr. Lwoff, médecin des asiles de la Seine vint passer
ses vacances avec sa famille. Le jeune André, suivant les sages conseils
de son père, avait préparé le S P C N; il y avait ajouté, en Sorbonne,
le certificat de Botanique; et, au Museum, les démonstrations de technique histologique données au Laboratoire d'Anatomie comparée.

En face de Carantec se trouve la pointe de Roscoff, et son laboratoire, fondé par Lacaze-Duthiers, où l'activation chimique de l'oeuf
d'Oursin avait été obtenue par Yves Delage. Passionné par tout ce qui
concerne le vivant, le jeune André y vint suivre les exercices pratiques
dirigés par Fred Vlès; il y revint l'année suivante pour suivre les
démonstrations données par Lucienne Dehorne, laquelle le reconnut
bientôt comme un étudiant hors série.

A ce même moment Félix Mesnil cherchait un assistant pour son
laboratoire de l'Institut Pasteur, et c'est avec une heureuse arrière-
pensée que Lucienne recommanda André à Edouard Chatton; celui-ci
mit aussitôt le jeune homme à l'épreuve en l'associant à ses propres
travaux.

Édouard Chatton avait déjà rencontré chez les Gondis et les
Damans du Sud-Tunisien quelques Ciliés parasites pour lesquels il créa
la famille nouvelle des Nicolellidae; sa perspicacité lui montra, dans ce
groupe, un curieux exemple d'évolution orthogénétique par migration
progressive de la zone buccale. A Strasbourg il avait abordé l'étude
expérimentale de la bipartition d'un *Glaucoma,* et obtenu, par un effet
d'inhibition, des individus à deux bouches. A Roscoff, il trouvait sur
les branchies des Acéphales, puis chez nombre d'organismes marins, de
très nombreuses espèces de Ciliés parasites, très insuffisamment connues,
et plus ou moins curieusement adaptées aux conditions offertes par leurs
hôtes. C'est avec André Lwoff qu'il en aborda l'étude, et cette collaboration fut un éclatant succès.

Nous avons déjà dit que Chatton était un travailleur infatigable,
sagace, tenace, obstinément appliqué à suivre pour le dénouer, le fil
embrouillé d'un cycle parasitaire. Tâche ardue et particulièrement difficile lorsque le parasite empruntait successivement des hôtes différents.

Les qualités d'André Lwoff, ses dons d'observateur, sa perspicacité, son opiniâtreté étaient à la hauteur des qualités de son maître.
L'efficacité de leur collaboration se solda aussitôt par de remarquables
trouvailles. Ce fut d'abord, en 1921, la description de quelques espèces

du genre nouveau *Sphenophrya;* point de départ d'un ensemble de recherches, de découvertes, d'études comparatives aboutissant plus tard à définir, dans une remarquable synthèse, l'ordre nouveau des Thigmotrichida. Ce fut en 1922 la découverte d'*Ellobiophrya donacis,* étonnant exemple d'adaptation structurale réalisée par un Cilié Peritrichida, déjà hautement différencié. Ce fut encore la découverte, en 1923, à Banyuls, du Cilié *Spirophrya subparasitica* qui devait conduire, par de nombreuses étapes, à la création de l'ordre nouveau des Apostomatida, puis au magistral exposé publié en 1935 et dédié à Felix Mesnil. L'avant propos de ce travail débute par ces mots où l'on retrouve le style et la pensée chattoniens: "Il y a un peu plus de douze ans, qu'à la Station biologique de Roscoff la collaboration des auteurs de ce travail s'est établie. Au moment où, en septembre 1924, il prit naissance au laboratoire Arago, à Banyuls-sur-Mer, elle était déjà si étroite, que ni l'un ni l'autre ne tenteraient de dire aujourd'hui, la part qu'il a prise aux observations premières et au long labeur qui s'en est suivi. Ce labeur a occupé toutes leurs vacances."

IV. OÙ IL SERA QUESTION DU "ZIZYPHINUS EXASPERATUS" GRAY

C'est au cours de la même époque que parfois, durant l'été, je prenais plaisir à quitter trois ou quatre jours le laboratoire du Croisic, ou, plus tard, celui de Concarneau, pour faire une virée à Roscoff; ce pourquoi je suivais quelque itinéraire fantaisiste, combinant l'usage de la bicyclette et celui du réseau à voie étroite des pittoresques chemins de fer bretons.

Arrivant à l'improviste à la Station biologique, j'y rencontrais des maîtres, des camarades, des amis, dans toute la fièvre du travail ou dans l'impatiente attente d'une prochaîne et favorable récolte, mais aussi dans cette euphorique et pleine liberté d'esprit que procurent le vent du large et l'absence des contraintes vestimentaires, protocolaires et mondaines. Je n'oublie pas tel soir, à la Maison Blanche, où, lorsque j'eus subi en présence des camarades et sous la présidence du prince Cantacuzene, les épreuves rituelles, j'eus l'honneur d'être proclamé Zèbre. Et je me souviens encore d'une curieuse constatation faite au matin suivant, lorsque, parti de belle heure (après avoir trouvé mon sac de voyage rempli de vieilles ampoules électriques brûlées!) je voulus adresser, de Morlaix, un cordial au-revoir postal aux copains: mon stylo ne voulut rien écrire, et je compris bientôt qu'il était parfaitement "drunk," étant plein non plus d'encre, mais de calvados!!

Charles Perez, Fred Vles, Édouard Chatton, Lucienne Dehorne, André Lwoff, Robert Courrier, René Wurmser, Boris Ephrussi, Louis Rapkine, Maurice Parat, Jean Cantacuzene, Albert Dalcq, Harold Munro Fox—je cite au hasard ces quelques noms associés à tant de bons souvenirs et je reviens à notre sujet.

A Roscoff, André Lwoff apparaissait avec la tranquille et élégante assurance que procurent la jeunesse, une stature honorable, un habitus viril et bien charpenté, et l'abondante chevelure dorée qui ajoutait à ses dons une marque évidente d'originalité. Il faut compléter cette esquisse en évoquant sa mise soignée et la distinction avec laquelle il portait négligemment une veste de velours demeurée célèbre.

Son affabilité, son attitude déférente à l'égard de ses maîtres et de ses aînés, témoignait de la sincère et saine modestie, habituelle à ceux qui, sans ignorer leur propre valeur, savent reconnaître celle des autres. Mais il faut ajouter que ses jugements sûrs et pénétrants s'exprimaient parfois, sans dissimulation, par un malicieux sourire d'euphorique ironie ou, plus durement, par quelque reflexion mordante.

Édouard Chatton était un réalisateur énergique, mais souvent ombrageux, inquiet, anxieux de savoir s'il avait suivi le bon chemin et trouvé la solution correcte du problème. André Lwoff était optimiste, rigoureux dans la recherche et parfois mystérieux dans ses attitudes. Abordant l'exposé d'une question difficile avec toute l'autorité que confère la constitution XY, il lui arrivait par moment d'esquisser une petite moue réticente, à la BB (avant la lettre), suivie, en relevant la tête, par un léger plissement des yeux, accompagné d'un sourire énigmatique laissant entendre qu'il y aurait encore beaucoup de choses à dire. . . .

Je dois à l'amitié d'Alexandre Cantacuzene, deux photographies de Roscoff faites par notre ami Jacques Millot qui a bien voulu me permettre de les utiliser.

Zizyphinus exasperatus Gray; je n'ai jamais su pour quelle raison roscovite le nom de ce joli Gasteropode à la coquille conique terminée

A mon ami
et complice
Dédé Lwoff.
EChott

Dédé observant
l'échappe des promelés
avec concupiscience
Banyuls. octbre 1925

en pointe aigue avait été associé, un jour, à notre jeune et brillant cama-
rade. Faut-il rappeler ici l'Histoire naturelle des Professeurs du Jardin
des Plantes publiée en 1847 par Isidore de Gosse avec la description
d'espèces nouvelles telles que *Ichthyologus affabilis* Gray, *Anatomicus
erinaceus* Linné, ou encore (mais ceci devient méchant), *Thuriferarius
Dumassianus* Oken dont les correspondants respectifs étaient MM. de
Lacepede, de Blainville, et Cahours?

V. OÙ L'ORGANISATION DES CILIÉS APPARAIT SOUS UN
JOUR NOUVEAU

Au cours de leurs premières recherches sur les Ciliés parasites, Chatton
et Lwoff ont utilisé la technique de Klein au nitrate d'argent malgré les
graves inconvénients dus à la dessication de l'objet étudié. De patients

et nombreux essais leur permirent bientôt de tourner la difficulté et d'obtenir des images entièrement satisfaisantes en enrobant les Infusoires correctement fixés au préalable, dans une mince couche de gélose ou de gélatine à travers laquelle le sel d'argent diffuse avant d'imprégner les corpuscules basaux des cils. L'infraciliature est dès lors visible dans son ensemble, sans déformation du corps, et sur les deux faces de celui-ci. Cet important résultat permettait de fonder sur des données précises la morphologie comparée des Infusoires étudiés, d'analyser les transformations de l'appareil ciliaire marquant les différents stades de leur cycle évolutif, et de déterminer les mécanismes assurant leur duplication au cours de la division. Une interprétation nouvelle des structures ciliaires devenait alors possible.

L'étude des Flagellés avait conduit Édouard Chatton à la notion de "cinetide", ce terme désignant l'unité structurale autoreproductible constituée par le blépharoplaste, le cil, les racines ciliaires et d'autres constituants annexes. Il avait substitué au terme blepharoplaste celui de cinétosome pour marquer, conformément au concept formulé par Henneguy et par v. Lenhossek, l'identité fonctionnelle de ce corpuscule et du centrosome.

Les rangées ciliaires méridiennes des Infusoires, ou "cinéties", pouvaient être considérées comme des séries de cinétides reliées par un cordon fibrillaire latéral, ou "cinétodesme", et la règle de "desmodexie", établie avec André Lwoff affirmait à la fois la structure dissymétrique et la polarité des cinéties, en même temps que la notion de division homothétigène affirmait leur individualité.

Ces notions nouvelles, fondées sur le prodigieux ensemble de faits accumulés par Chatton et Lwoff, renouvelait entièrement l'interprétation des ciliatures, celle des appareils ou des systèmes ciliaires spécialisés et plus particulièrement encore, celle des mécanismes assurant, au cours de la bipartition, la réorganisation, chez les deux individus fils, de la ciliature caractéristique de l'espèce.

Au cours d'un exposé général, J. O. Corliss retient comme les plus importantes des "Chattonian ideas . . . the autonomy and genetic continuity of the Kinety, the pluripotency of kinetosomes, the rule of desmodexy, the significance of stomatogenesis, the development of the concepts of homothetigenic and symmetrogenic modes of fission".

C'est pourquoi les recherches méthodiquement poursuivies par Chatton et Lwoff au cours d'une série d'étés marquent une période dans l'histoire de la Protistologie.

VI. OÙ L'AUTEUR DE CES LIGNES REVIENT À SES PREMIÈRES AMOURS

Les satisfactions que m'avait procuré l'oeuf de *Parascaris equorum* n'eurent pas de lendemain. Les curieux processus biochimiques que des techniques très simples m'avaient permis de mettre en évidence n'avaient, contrairement à mes premiers espoirs, ni valeur générale, ni signification généralisable; ils se rapportaient essentiellement à la formation des membranes protectrices de cet oeuf, c'est à dire à des mécanismes biologiques secondaires et, de plus, très particuliers. De nouvelles recherches, portant sur l'oeuf de l'Annélide *Sabellaria,* n'apportèrent que peu de données nouvelles, en marge des problèmes essentiels de la vie cellulaire, que les méthodes utilisées ne permettaient pas d'atteindre. Restant préoccupé par l'aspect moléculaire des structures organisées, j'abordais alors l'étude des lames minces protoplasmiques et celle des changements d'état physique liés à l'activité des cellules macrophages. Enfin, j'espérais trouver dans la structure macromoléculaire des fibres protéiques telles que collagène ou élastoïdine, un modèle simple, maniable, accessible à l'expérimentation, et permettant d'aborder l'interprétation physico-chimique de l'organisation.

Avant d'avoir compris qu'au seul point de vue technique ces tentatives étaient prématurées, mon travail fut interrompu brutalement par les évènements de 1939 et supplanté par d'autres occupations.

Revenu au Collège de France, l'année 1940 me posa un problème; il fallait trouver un objet d'étude que l'on puisse se procurer aisément, à bon compte, en toute saison, sans avoir à solliciter quelque permission des autorités du moment, et que l'on puisse exploiter à peu de frais. Les Protozoaires en général, et les Infusoires Ciliés en particulier, répondaient à ces voeux.

Les Ciliés, à eux seuls, constituent un monde inépuisable. Les Epinoches d'un ruisseau nogentais m'apportèrent deux Infusoires réalisant chacun par des moyens différents un anneau de fixation comparable à celui d'*Ellobiophrya*. Ils furent aussitôt baptisés, l'un: *Erastophrya Chattoni,* l'autre *Epistylis Lwoffi.* André Lwoff était alors à l'Institut Pasteur et Edouard Chatton à Banyuls d'où je reçus en date du 13. I. 1944 les mots suivants:

"Le laboratoire a été évacué de fond en comble il y a bientôt deux mois. Nous campons scientifiquement et végétativement en attendant un nouvel épisode.

C'est alors une vraie joie que de recevoir le message impersonnel qui évoque cependant le mieux tout ce qu'il y a entre l'expéditeur et le destinataire d'affinités profondes et d'amitié bien et longuement vécue. Lwoff m'avait touché deux mots de vos deux récentes découvertes et m'avait dit que vous consacriez notre complicité en nous les dédiant simultanément. Vous ne pouviez pas nous faire plus de plaisir et cependant l'envoi que vous me faites de votre note y ajoute encore, surtout dans le temps présent, une teinte d'union et de confiance en l'avenir qui m'est très précieuse".

Reprenant ainsi l'étude des Infusoires Ciliés, avec Marie Hamard tout d'abord, le guide indispensable se trouvait être la technique d'observation et les notions morphologiques données par Chatton et Lwoff. Mais, si ces derniers ont su exploiter avec un succès sans égal, la manie pastorienne des espèces parasites, il faut reconnaître que le monde des ciliés libres offre un champ de recherche illimité. C'est avec la collaboration de Janine Ducornet, d'Opaline Mugard, de Michel Tuffrau, de Jean Dragesco, de John Corliss qu'il a été possible d'en étendre l'exploration, puis d'esquisser une révision générale de la classe des Ciliata.

VII. OÙ LE MOMENT EST VENU D'ÊTRE BREF

Au laboratoire de l'Institut Pasteur, André Lwoff trouvait ample matière à réflexion sur les multiples problèmes posés par la physiologie des microorganismes. Considérant la diversité de leurs modes de croissance, de leurs besoins nutritifs, des conditions de leur culture *in vitro,* et encore les pertes de fonctions subies au cours d'une longue évolution différenciatrice, il réalisait progressivement, dans ce large domaine, une des parties les plus originales de son oeuvre. Ce que d'autres exposent mieux que je ne le saurais faire, dans les pages adjacentes.

Je rappellerai cependant que Lwoff fut le premier à obtenir la culture axénique en milieu liquide peptoné, d'un Cilié ordinairement microphage, *Tetrahymena riformis,* inaugurant ainsi la longue série de recherches expérimentales poursuivies par divers auteurs sur le métabolisme de cet Infusoire.

Dans un autre domaine, c'est après avoir efficacement confirmé le concept chattonien du cinétosome-organite cellulaire autoreproductible, que André Lwoff, élargissant le problème, organisait en 1949 un Symposium consacré au problème des unités biologiques douées de *continuité génétique.* Mais cette notion profonde de continuité génétique, changeait progressivement d'aspect en s'adaptant à des cas nouveaux. Le cycle d'un Bactériophage, la découverte de la lysogénie, le concept de prophage, telle est la magnifique contribution lwoffienne aux débuts de la Biologie moléculaire, cette Science des mécanismes intimes de la vie, aux progrès de laquelle notre ami ne cesse de collaborer.

Pendant ce temps, l'usage du microscope électronique, que Édouard Chatton n'a pratiquement pas connu, s'est imposé dans tous les domaines de la Cytologie. Les images obtenues ont profondément transformé l'interprétation des structures cellulaires. Les Protistes en général et les Ciliés en particulier n'ont pas échappé à l'emprise de ce nouveau moyen d'investigation qui nous montre—enfin—la structure d'un cinétosome et comment cette structure peut induire, dans son très proche voisinage, la cristallisation d'une autre structure semblable. Est-ce à dire que tout est à recommencer? Pas tout à fait, certes, mais une profonde revision de nos connaissances morphologique sur les Ciliés est en cours. Et c'est une joie sans mélange que de voir revivre, sous un jour entièrement nouveau les notions que notre jeunesse avait entrevues et les faits dont elle croyait avoir réalisé la conquête.

Aristotle–totle–totle

M. DELBRÜCK

THE LOGICIAN: *Let us take another example. All cats are mortal. Socrates is mortal. Therefore, Socrates is a cat.*
THE OLD GENTLEMAN: *And has four feet. Indeed, I do have a cat named Socrates.*
THE LOGICIAN: *There, you see. . . .*
THE OLD GENTLEMAN: *Then Socrates really was a cat!*
THE LOGICIAN: *That is what Logic reveals to us!*

E. IONESCO, *The Rhinoceros*

During the past 20 years André Lwoff—*mon cher collègue et ami*—and I have been carrying on a casual correspondence about Aristotle, inspired by a remark in a public lecture by another dear friend and colleague, who claimed that Aristotle, *more than 3000 years ago,* had said such and such a thing about Life. It stands to reason that a prenatal quote from Aristotle, centuries before his birth, is something special and precious, whatever its content. Dr. Lwoff and I, therefore, endeavored to find other, possibly still earlier utterances that might shed light on the intellectual development of the great sage from Stagira. And we did not labor in vain. Indeed, in one of Dr. Lwoff's most recent letters to me he was able to communicate a quotation dating back more than 4000 years, adding, justly, "The more you push him into the remote past the more impressive the man becomes."

Unfortunately, I am not in a position to publish here these metahistorical studies, since this would have required the consent of Dr. Lwoff. The rules set up by the editors of this book explicitly forbade such an approach. Thus it must be left to future historians of science to dig out the fruits of our labors from the appropriate archives. As far as Dr. Lwoff's letters to me are concerned, I can throw out the hint that they will be deposited in the Archive of the Millikan Library of the California Institute of Technology.

My letters to Dr. Lwoff should be in his files. I must confess that I tried to obtain copies of these letters by stealth. I wrote to Jacques Monod, suggesting that Gisèle might be able to find them. He expressed enthusiasm and vowed cooperation. But, as we all know, Jacques is undependable. "At lovers' perjuries, they say, Jove laughs." He laughs at Jacques's perjuries, too. I never heard from him again.

Nevertheless, while we are talking about Aristotle, I should like to utilize this opportunity to state the conjecture that this wonderful man discovered DNA. Let me explain.

To consider Aristotle not as a philosophical system but as a human being subject to development is an idea of this, our twentieth century. Werner Jaeger (1925) was the first to pursue this approach, with a vengeance, thereby ushering in a new era of Aristotelian studies. Now it so happened that Jaeger knew next to nothing about biology. He lived before the double helix had hit the news and could not see anything of interest in either the biology of his own day or in Aristotle's five major biological books (*Historia animalium, De partibus animalium, De motu animalium, De incessu animalium, De generatione animalium*). In fact he considered these books as something compiled by Aristotle in his old age, illustrative applications of his general views on natural philosophy and metaphysics. Scrutiny of the internal evidence by a host of later students has modified this view to the extent of placing some of these books in the period of Aristotle's travels with Theophrastos on Lesbos and in Macedonia, before his return to Athens and the founding of his own school, but definitely after his formative period of 20 years in Plato's Academy. Nobody can fail to be impressed with the wealth of biological observations, and Ingmar Düring (1965) points out the intensity, variety, and subtlety of the speculative arguments. He, too, however, puts these studies after the main philosophical opus, and especially attributes *De generatione animalium* to Aristotle's late period.

This chronology may well be correct for the books we have. However, I would like to conjecture (and I would not assume that I am the first to do so, having assimilated only 10^{-3} of the relevant literature) that biological studies constituted the decisive early intellectual experience in Aristotle's life, imprinting *telos* on him as his most pervasive concept. Where Plato sees the world as ideas of which static objects are the shadows, Aristotle sees striving and development towards goals and motions governed by permanent plans.

The following passage from *De partibus animalium* (I, 5; 644b21–645a37) contrasts the eternal world of astronomy with the seemingly ephemeral one that surrounds us on Earth:

Of the products of Nature some are eternal, not subject to generation and corruption, others grow and perish. Of the former, grand and divine as they are, we have less insight since they offer few aspects for our perception. From these scanty data we can explore what we care to know about them. In contrast, for the perishable things, plants and animals, we are given a great wealth of information since they are close at hand. If one makes the effort, much can be learned about each kind. Both sciences have their charm. Even though our understanding of the eternal things is more limited, they fascinate us more than the things of our own world because of their grandeur, just as our imagination gets more excited by even a glimpse of a beloved person than by the close observation of many other and even important things. However, the perishable things are to be preferred as objects of science because of the wealth of knowledge we can gather about them. I will speak about the nature of animals and to the best of my ability not overlook anything, may it seem large or small. Also with those less appealing creatures, nature grants indescribable pleasures to those with a scientific bent, by revealing her creative power to their scientific scrutiny. Indeed it would be absurd were we to take delight in artistic reproductions, admiring the craft of the artist—as we do with paintings and sculptures—and should not take delight in the original creations of nature, especially when we can achieve some measure of understanding of their structure. Therefore one should not childishly recoil from the study of lower animals. All creations of nature are miraculous. When strangers were visiting Heraklitos and found him warming himself by the oven they hesitated to enter. He encouraged them to approach, saying, "The Gods are here, too." Just so one should approach the study of any animal with reverence, in the certainty that any of them are natural and beautiful. I say "beautiful" because in the works of nature and precisely in them there is always a *plan* and nothing accidental. The full realization of the plan, however, that for which a thing exists and towards which it has developed, is its essential beauty. Also one should have it clearly in mind that one is not studying an organ or a vessel for its own sake but for the sake of the functional whole. One deals with a house, not with bricks, loam, or wood. Thus the natural scientist deals with the functional whole, not with its parts, which as separate entities have no existence.

This famous passage could be entitled "A Biologist Looks (Somewhat Defensively) at Physics," and it is not the only one in which Aristotle is anxious to point out that the world of creatures has its kind of eternity, too. In *De generatione animalium* we find (II, 1; 731b, 32–39) this sentence: "Since it is impossible that creatures should be eternal, these things which are generated are not eternal as individuals (though the essence is in the individual) but as a species."

Anybody who is familiar with today's physics and biology, and who reads Aristotle's writings in these two fields, must be struck by the aptness of many of his biological concepts, in contrast to the tangled inconsistencies of his physical and cosmological analyses. And, indeed, nobody would deny that Aristotle's physics was a catastrophe, while his biology abounds in aggressive speculative analysis of vast observations on morphology, anatomy, systematics, and, most importantly, on embryology and development.

Aristotle does consider it remarkable, and a fundamental aspect of Nature, that human beings beget human beings, and do not beget rabbits or an ear of corn.[1] What strikes the modern reader most forcibly is his insistence that in the generation of animals the male contributes, in the semen, a *form principle,* not a mini-man. He argues, contra Hippocrates, that the semen is *not* a secretion, in which each part of the body is represented by a contribution from that part, pointing out:

(a) The resemblance of children to parents is no proof of part-for-part representation because the resemblance is also found in voice and in way of moving (GA I, 18; 722a, 4–7).

(b) Men generate progeny before they have certain parts, such as beards or gray hair (722a, 8–9); similarly with plants (722a, 12–14).

(c) Inheritance may skip generations "as in the case of the woman in Elis who had intercourse with an Ethiopian. Her daughter was not dark but the daughter's son was" (722a, 10–12).

---

[1] See (and hear) the 5-minute lecture (with *guitarre*) on molecular genetics by Joel Herskowitz, entitled "The Double Talking Helix Blues," a phonograph record published by The Vertebral Disc, 913 S. Claremont, Chicago, Illinois 60612.

(d) Since semen can determine female children, it clearly cannot do so by being a secretion, in a man, from female genitals (723a, 31–32).

From the foregoing it is clear that the semen does not consist of contributions of all parts of the body of the male (as Hippocrates had taught), and that the female's contribution is quite different from the male's. The male contributes *the plan of the development* and the female the substrate. For this reason the female does not produce offspring by herself, since the form principle is missing, i.e., something to begin the development of the embryo, something that will determine the form it has to assume (GA I, 21; 730a, 24–30).

The form principle is likened to a carpenter. The carpenter is a moving force which changes the substrate, but the moving force is not materially contained in the finished product.

The semen contributes nothing to the material body of the embryo but only communicates its program of development. This capability is that which acts and creates, while the material which receives this instruction and is shaped by it is the undischarged residue of the menstrual fluid (GA I, 21; 729b, 5–8).

The creature produced from them (the form principle in the semen and the matter coming from the female) is produced like a bed comes into being from the carpenter and the wood (729b, 17–18).

The male contributes the principle of development, the female the material (730a, 28).

The male emits semen in some animals and where he does, it does not become part of the embryo; just as no part of the carpenter enters into the wood in which he works, . . . but the form is imparted by him to the material by means of the changes which he effects. . . . It is his information that controls the motion of his hands (GA I, 22; 730b, 10–19).

Quite a few quotations in a similar vein could be added. Put into modern language, what all of these quotations say is this: The form principle is the information which is stored in the semen. After fertilization it is read out in a preprogrammed way; the readout alters the matter upon which it acts, but it does not alter the stored information, which is not, properly speaking, part of the finished product. In other words, if that committee in Stockholm, which has the unenviable task each year of pointing out the most creative scientists, had the liberty of giving awards posthumously, I think they should consider Aristotle for the discovery of the principle implied in DNA. It is my contention that Aristotle's principle of the "unmoved mover" originated with his biological studies, and that it was grafted, from here, first onto physics, then onto astronomy, and finally onto his cosmological theology.

I should like to suggest, furthermore, that the reason for the lack of appreciation, among scientists, of Aristotle's scheme lies in our having been blinded for 300 years by the Newtonian view of the world. So much so, that anybody who held that the mover had to be in contact with the moved and talked about an "unmoved mover" collided head-on with Newton's dictum: Action equals reaction. Any statement in conflict with this axiom of Newtonian dynamics could only appear to be muddled nonsense, a leftover from a benighted, prescientific past. And yet, "unmoved mover" perfectly describes DNA: it acts, creates form and development, and is not changed in the process.

Indeed, let us go one step further while we are in the mood, *mes très chers collègues et néanmoins mes amis,* and consider the fact that the re-entry of Aristotle into Western thought occurred through scholastic Christian theology. Let us assert that, by the irony of history, the vast historical impact of Aristotle on Western thought is the result of an almost accidental appropriation of the most secondary and misguided aspects of Aristotle's speculations, and that it is due to this bizarre twist that we are encumbered today with a total barrier of understanding between the scientist and the theologian, from St. Thomas Aquinas till today, Catholic, Protestant, and LSD mystic alike. Thus a new look at Aristotle the biologist may yet lead to a clearer understanding of the concepts of purpose, truth, and revelation, and perhaps even to something better than mere coexistence between us natural scientists and our colleagues from the other faculties.

REFERENCES

Düring, I. (1966). "Aristoteles. Darstellung und Interpretation seines Denkens," Heidelberg, Carl Winter Universitätsverlag.
Jaeger, W. (1923). "Aristoteles. Grundlegung einer Geschichte seiner Entwicklung," Berlin; *Nachdruck, 1955.*

A latter-day
rationalist's lament

⤔⤳

S. E. LURIA

*Oubliez pour un moment le point que nous occupons dans l'espace
et dans la durée, et étendons notre vue sur les siècles à venir, les
régions les plus éloignées et les peuples à naître. Songeons au bien
de notre espèce; si nous ne sommes pas assez généreux, pardonnons
au moins à la nature d'avoir été plus sage que nous.*

DIDEROT, *Le Neveu de Rameau*

In planning this essay in honor of André Lwoff, I discarded one after
the other a series of topics that came to my mind: lysogeny, microbiology and man, my early and more recent days at the Institut Pasteur.
At the moment, thoughts of André Lwoff are associated in my mind
with more urgent concerns about the future of our culture and civili-
zation.

Perhaps I can explain what I mean by recalling how Lwoff punctu-
ated a classical review on "Lysogeny" (1), as well as other writings,
with quotations from Aeschylus, Bacon, Poincaré, Nietzsche, and Gide,
among others—quotations sometimes seriously apt, sometimes whimsi-
cal, all revealing a concern for intellectual clarity and an urge to relate
scientific ideas to the whole of a cultural tradition. André Lwoff—
scientist, painter, master of language, leader of one of the great schools
of biology—is a prototype scientist-humanist, in whom the "two cultures," supposedly divergent and losing touch of each other, remain

happily married. In a true sense, such a person is the living embodiment of the cultural tradition of modern man, in which the rationalist-scientific enterprise of the last three centuries has been grafted on the vigorous stem of humanistic culture.

But today that tradition is in crisis. Its value for man and its future viability are being challenged, especially within those societies that have felt for the longest time the impact of its consequences. The challenges are of different but not unrelated kinds. On the one hand, the scientific-rationalist culture is challenged by those who see it as a sterile, mechanical, value-less pursuit of knowledge, blind to the consequences that this knowledge may have in store for humanity or, even worse, as a handmaiden to technocracy (2). On the other hand, there is the challenge of those who see the rationalist-scientific culture as a transient phenomenon, the product of special local conditions in Western countries in the last few centuries, and one that has nearly reached saturation, so that its presumed goals—knowledge and progress—are coming to an asymptotic end (3).

Both challenges reflect the tragedy of our times. An enormous amount of scientific progress has generated stupendous technologies; but the rationalist hopes of the past two centuries, that a free-wheeling technology would satisfy the biological needs of man, remove the causes of conflict among men, and open up a royal way to a golden era of humanity, have proved illusory. The brutalities of two world wars at a time when educated men considered war unthinkable; the rise of aggressive nationalisms just when the way seemed open to international brotherhood; the persistence and increase of poverty even in the most affluent societies; and the current twin threats of nuclear destruction and uncontrolled overpopulation—all these events have shattered confidence in the values of the rationalist-scientific revolution.

The whole enterprise is being rejected as responsible for, and irresponsive to, the problems of human life. Increasingly men, especially young men, search eagerly for alternative sources of values: either turning to traditional religion, or sinking into solipsistic despair, or seeking mystical solutions that substitute immediate experience, individual or collective, spontaneous or drug-enhanced, for logical analysis. The ultimate aberration—the rejection of sanity itself—is manifested in claims of the positive value of psychotic experiences (4)!

What has gone wrong, and what is to be done? I believe the root of the problem to be, not in the scientific direction of the cultural enterprise, but in its uneven development and in the conditions of the society in which its products are being applied. Science produces knowledge, which becomes the source of technology. Technology in turn is being used for social and political goals. But the society that uses technology is not a rational society. Its goals are not chosen rationally, in a way that would bring about the maximum satisfaction of human needs and desires. Hence most uses of scientific technology fail to benefit mankind. At best, they satisfy some short-range demand; at worst, they serve the profit or power motives of ruling groups.

What is wrong is not a surfeit of science and technology, but their uses in a society that has not achieved means to choose its own goals rationally. The need is not for alternatives to science, but for a science of man and of its society, a science of human technology, and a rational approach to human values.

Man is a unique animal. He has a biological nature and also, uniquely, a cultural nature. The biological nature makes for the transience of each individual; the cultural nature, based on language, makes for the persistence of individual contributions. I dare to assert that these two natures of man contribute to humanity as a whole its only rationally conceivable "goals": biological survival and cultural development. The first is Darwinian, the second Lamarckian. Together, these two impersonal goals confer meaning to the existentially purposeless life of the individual human beings. Ultimately, these collective goals are the only rationally acceptable sources of meaningful, evolving values by which humanity can rule itself at any one point in its development.

But the two goals do not necessarily function in a cooperative way. In fact, they are often in conflict. Thus at some times the cultural developments may pose a threat to man's biological survival; we may be at such a point now with regard to overpopulation and nuclear energy. At other times, man's biological needs such as those that generate mass human migrations may play havoc with some cultures, as took place in the dark ages of post-Roman Europe.

At such times, remedies are needed to restore a balance and enhance man's survival and cultural development. But what are those remedies? Some frightening perspectives have been conceived by literary imagination: for example, the biologically manipulated, culturally immobilized *Brave New World* of Aldous Huxley. Hardly less frightening is the society conceived by Hermann Hesse in *Magister Ludi,* in which culture becomes an esthetic cult delegated to a small, uninfluential elite.

More significant as a remedy, if only a palliative one, is the quest, forcefully advocated by many in our society, that man's cultural activities be validated by the criterion of immediate relevance to social problems. Scientist and humanist are taken to task for their insulation from the problems of society, for their presumed pursuit of sterile intellectual exercises, for widening the gap between the amount of knowledge available and the social environment in which that knowledge is being applied. There is some validity in this demand for relevance. A wide gap exists between the cultural enterprises that modern societies encourage and finance and the social technologies needed to repair and prevent the ills that exist within these societies.

Yet, when the quest for relevance in scientific and cultural activities takes an anti-intellectual, anti-rationalist turn, its aim is wide of the true target. It is not culture that has failed society, but society that is failing culture by applying culture's output to goals that do not make for man's success, either biological or cultural.

If the two classes of human goals, biological survival and cultural development, are to flourish together, not less but more science may be needed. We need a science of human nature that can provide an understanding of the needs, drives, and interactions of men with each other and with their environment (5). Today, because of misguided priorities, we may know more about man in interplanetary space than about men on earth in their own rural or urban dwellings, about their reactions to close relatives or to alien strangers, to people of similar or different color or language.

Also, we need a social science of technology that studies the impact of new technologies on society and can foresee the probable results and their social consequences (6).

More than anything, we need a political science that takes into account the biological and cultural nature of man and devises political structures that enhance the dual goal of humanity rather than the narrow interests of a class or a nation, or the fostering of some fixed ideology. Social science has lagged behind natural science, not only because of the intrinsic difficulties of the subject, but because of the prejudices and taboos it has encountered, of the resulting relative lack of encouragement and support, and of having too often let itself be made the faithful servant of the status quo. Some of the sharpest insights into the possible methodology of a scientific social science—by Marx and Freud, for example—have not yet flourished in the same fashion as the great insights of natural science because of being or becoming embedded in rigid or metaphysical frameworks.

How can society develop the kind of social science that can discover, formulate, and forcefully proclaim sets of values that reflect and respect the dual goals of humanity at any one stage of its development? How can we insure that the products of such a social science, when available, will be implemented in a rational, dynamic, evolutionary way? And, most important of all, is there time left? The biological crisis of humanity is drawing near; the cultural crisis may already be at hand, both results of the uneven development of different branches of human culture.

I for one am unwilling to despair, if only because the basis for hope is deeply embedded in my personal allegiance—as a biologist, to man's biological survival; as an intellectual, to man's cultural progress; as a rationalist, to the rationalistic direction of that progress in the future. I believe that the outdated frameworks of beliefs and the stultifying power relations of present-day societies can be made to give way to a more adaptive set of social relations and political institutions. I believe that if the passion, the methodology, and especially the commitment to factual truth and mutual trust characteristic of the natural sciences can be applied to developing a humanistic social science and implementing its findings in the social arena, mankind will find a path in which its dual goals will be pursuable together in harmoniously balanced relation. In that path, the quest and reverence for knowledge, not only of the physical world, but of man and his unique passage in the universe, may flourish jointly with the recognition and respect for his biological nature.

NOTES

1. André Lwoff, Lysogeny, *Bact. Rev.* 17, 269–337 (1953).

2. See discussion by T. Roszak, *The Making of a Counter Culture,* Doubleday, New York, 1969.

3. See, for example: G. S. Stent, *The Coming of the Golden Age,* Natural History Press, Garden City, New York, 1969.

4. See, for example: *The Value of Psychotic Experience,* a summer series of workshops and symposia presented by Esalen Institute, Big Sur, California. See also R. D. Laing, *The Politics of Experience,* Pantheon, New York, 1967.

5. For a penetrating discussion by a biologist, see H. Gaffron, Resistance to Knowledge, *Ann. Rev. Plant Physiol.,* 20: 1–40 (1969).

6. See P. Goodman, Can technology be humane? *The N.Y. Rev. Books,* 13 (No. 9), 27–34 (1969).

"When a thing's really good it cannot die"*

❦

HILARY KOPROWSKI

BY WAY OF A DEDICATION

We are returning from a concert of the Philadelphia Orchestra. Kertesz conducted "Les Francs-Juges Overture" by Berlioz and one of the early Mozart symphonies: No. 35, K. 319. André is enjoying the music. After the intermission, Dvořák's Symphony No. 7 is presented. André remarks, "I am not a musician, but Dvořák's Symphony sounds rather disjointed to me; there is no central theme, no leading musical thought." From a "nonmusician," this is a perceptive critique of the loosely assembled folk tunes and emotional outbursts which Dvořák assembled under the guise of a symphony.

This is the year 1969, the year of the four-letter word. André paints, I play the piano. We are looking for an island, for Candide's garden, but it is only talk. In his case, though, the creative impulse which made him do wonders in science may really be equally as powerful when applied to his canvases.

The camera turns back 22 years to the first postwar International Congress of Microbiology in Copenhagen. The brash, cocksure, well-fed scientists of the New World meet, for the first time since the war, their

---

* This title and the subtitles following, unless otherwise indicated, are taken from the song "Some Like It Hot," lyrics by Frank Llesser and music by Gene Krupa and Remo Brondi. © Paramount Music Corporation.

tired colleagues of the Old World. But on encounter, we find ourselves less brash, less sure—the Europeans lack material means but not brains. A curious amalgam of camaraderie between the two worlds develops against the background of Danish rationing. Why André was already a legend for me I cannot explain now. We are flying to Paris in the same plane. We sit together. Timorously I ask him what I can do for him. "Can you get from Jukes at Lederle some aminopterin and folic acid for me?"

This is the first of many encounters spanning the next 22 years. Some are brief and rather impersonal—at scientific meetings and parties; some more intimate and more intense, such as at our homes, or when, after seeing my mother at Nice, I drove to Grasse to spend several wonderful hours with André and Marguerite during their vacation.

"Il y a toujours la manière." Very true. Yes. There is the manner. The manner in laughter, in tears, in irony, in indignations and enthusiasms, in judgments—and even in love. The manner in which, as in the features and character of a human face, the inner truth is foreshadowed for those who know how to look at their kind.[1]

BY WAY OF FACTS

"WHAT A KILL-A-DILL-A!"

In "the house that Jack built," it was not the cat that "killed the rat," but heat. On April 4, 1968, the night of Martin Luther King's assassination, the "room" temperature in The Wistar Institute's Animal House started to rise at an alarming rate, and by 1:30 A.M. it reached 106°F (41°C). A jolly Finnish scientist, Pekka Häyry, at the time when "The moon shines bright, the stars give a light, and you may kiss a pretty girl at ten o'clock at night,"[2] sounded the alarm and started to smash windows in order to let in the cool night air.

In view of the earlier events of this day, the resulting commotion

[1] Joseph Conrad, *The Mirror of the Sea and a Personal Record:* "A Familiar Preface," London, J. M. Dent and Sons, 1946, p. xix.

[2] W. S. Baring-Gould and C. Baring-Gould, eds., *The Annotated Mother Goose,* New York, Bramhall House, 1962, p. 26.

was first taken for a riot. Only after the police were assured that it was only one man's fight for cool air, and not a mass outrage against society, did they let others in. A terrible sight awaited. The cool air had come too late. Unable to escape the shimmering, suffocating heat, most of the rats and the mice lay dead or dying.

"EVERYWHERE YOU CAN FIND SURVIVORS"

Contrary to the subtitle, survivors were few, except among the inhabitants of one animal room. This room housed C3HRv mice, a strain created at The Wistar Institute by the genetic transfer of a gene of resistance to arbo B virus from the Princeton-Rockefeller Institute (PRI) strain (Sabin, 1952) onto the arbo B virus-susceptible strain, C3H/He (subsequently referred to as C3H) (Goodman and Koprowski, 1962; Gröschel and Koprowski, 1965). Forty-seven out of 575 or 8% of C3HRv mice died from thermal shock, as compared to 72 out of 124 or 58% of the congenic C3H mice, which were susceptible to arbo B. The two strains were kept on the same shelves and exposed to the same intensity of heat. Because of the number of animals involved, the difference in their respective resistances to thermal shock was significant. In order to verify whether there could be any correlation between resistance to arbo B viruses and resistance to heat, we checked two other breeds of mice kept at The Wistar Institute, virus-resistant BRVR and virus-susceptible BSVS. These strains of mice were inbred (by Webster, 1937) but not congenic. Seven out of 50 or 14% of the BRVR mice died as a result of heat shock, as compared to the death of 100% of the BSVS mice.

"LONG AGO SWUNG A YOUNG GORILLA THRU THE TREES"

The mechanism of genetic resistance to infection is a subject dear to my heart. I have investigated this problem with several capable associates for more than a decade. We now know that resistance to arbo B virus, observed either on the organism or the cellular level, is interferon-mediated (Hanson and Koprowski, 1969) because the resistant animals and their cells are much more susceptible to the inhibitory action of preformed interferon on arbo B viruses than are susceptible animals. We still, however, have no clue as to the nature of the molecular events which would explain this strange phenomenon. Actually, at present, we are not even certain whether the phenomenon to be explained is that of *resistance* to the effect of interferon in *susceptible* mice or of susceptibility to interferon in the *resistant* mice. "He who has a choice has trouble," says a Dutch proverb; therefore, in the hope of simplifying our task rather than complicating it, we pounced upon the difference in heat susceptibility between the two mice strains as another possible approach to the problem of innate resistance.

For this reason, we wanted very much to confirm our original observations or to have them confirmed by someone else. We realized, however, that although "creating a disaster" may be valid from a statistical viewpoint, it certainly is not justifiable on either humanitarian or financial grounds. We therefore had to devise an approach to the heat resistance problem which would be more scientific in nature and less costly in terms of animal loss.

Another Finn, Kari Lagerspetz, Professor of Zoophysiology at the University of Turku and an authority on thermal regulation, was consulted, and he joined forces with Margo Darnell,[3] who investigates resistance in my laboratory. They designed experiments to determine the thermal death point of the mouse strains in question. The animals were maintained in an incubator, and the temperature was raised stepwise in three experiments. The results (Table 1) of these experiments confirm

TABLE 1. "SOME LIKE IT HOT; SOME LIKE IT COLD"

Experiment no.	Strain of mice	Gene of resistance to arbo B	No. of mice observed	Temperature at which all died (°C)	Duration of exposure (hours)
1*	C3H	Absent	6	41.3	4
	C3HRv	Present	6	43.5	6½
2†	BRVR	Present	9	42.1	8½
	C3HRv	Present	6	43.3	9½
3†	C3H	Absent	5	40.5	9
	BRVR	Present	6	42.0	10½
	C3HRv	Present	6	43.5	12

* Incubator temperature increased by 1° every hour.
† Incubator temperature increased by ½° every hour.

[3] A detailed account of these investigations will be published elsewhere by Lagerspetz et al.

the findings made after the disaster—the two strains of arbo B-resistant mice withstand thermal shock much better than do the arbo B-susceptible C3H mice.

"THANKS TO THOSE SENDERS AND ARRIVERS"

Kari Lagerspetz then took the mice involved in this study to Turku to investigate the mechanism of heat resistance in the C3HRv mice. He wished to determine whether the thermoregulatory mechanism functioned more efficiently in these mice because of better peripheral vascular control or because of better central control.

The animals were injected with noradrenaline for increase of their heat load. After comparing the colonic temperature readings with tarsal surface (skin) temperature readings, Lagerspetz concluded that the heat dissipation is managed more effectively in C3HRv than in C3H mice. There still remained, however, the problem of peripheral versus central control of thermoregulation in C3HRv mice. The mice were exposed to cold, and their heat retention capabilities were compared. The C3HRv mice, indeed, retained heat better than did the C3H mice, indicating that the vascular control mechanism in the virus-resistant mice functions more efficiently than that in the virus-susceptible mice, when the mice are exposed either to excessive heat or excessive cold. Thus, the central, hypothalamic thermoregulatory control mechanism is involved.

"NO, NO, NO, NO, NO; I SHOULD SAY NOT"

The next series of experiments was a valiant attempt to find a link between arbo B virus resistance and the more efficient thermoregulation in C3HRv mice. Both C3HRv and C3H mice were infected with the arbo B viruses, the West Nile virus, and the 17D strain of yellow fever virus, and then divided into two groups, one kept at room temperature (23°C) and the other at 35°C. We recorded the colonic temperatures frequently and eagerly awaited the outcome of the experiment. Our expectations, alas unfulfilled, were that the C3HRv mice would respond to virus infection (since the arbo B virus does replicate in their brain tissue) with a higher temperature, which could be controlled by their efficient thermoregulatory system and which would, in turn, combat the effect of the virus lethal for C3H mice. "Alas, alas, alas," as André would say, "the man in the moon came down too soon."[4] No differences were found in the temperature responses of the two strains of mice during the first 3 days after virus infection. Neither was there a difference between the temperature response of the C3H mice which were to survive the infection and that of those which were to die.

"EVER SINCE, THANKS TO MR. NERO, SOME LIKE IT HOT"

More interesting information was related to the discovery that a marked hypothermia was closely associated with the appearance of signs of illness in mice which were kept at room temperature. In these mice, progressive hypothermia always developed within 12–48 hours before their death. If these mice were kept at 35°C, hyperthermia and not hypothermia accompanied the development of illness, preceding death by 12 hours. Thus, although neither heat loss nor heat gain could account for the fatal outcome of infection, possible damage to the thermoregulatory center in the hypothalamus may be implicated.

The thermoregulatory disturbances as manifested by hypothermia or hyperthermia were much less frequently observed in C3HRv mice, even in the few which died after intracerebral West Nile infection, than in C3H mice. It is, therefore, possible to consider at present (probably to repudiate the notion later) that greater resistance to damage of the thermoregulatory center by virus infection may account for the greater resistance of the C3HRv mice to arbo B viruses.

"NON OMNES POSSUMUS ET NONULLI NOSTRUM NEQUEUNT"[5]

Within the context of the sheer speculative nature of the hypothesis of the mechanism of resistance, there are still many puzzles to be solved before we will see the light. For example:

1. Is there a gene for heat resistance and is it dominant, as in the case of the gene for arbo B resistance?

2. Can the heat resistance be expressed on the cellular level, similarly to the virus resistance?

[4] Baring-Gould and Baring-Gould, *The Annotated Mother Goose*, p. 83.

[5] A. A. Milnei, *Winnie Ille Pu* (translated by Alexander Lenard), Novi Eboraci: Sumptibus Duttonis, MCMLX, p. 51.

3. Was it possible while transferring the gene of arbo B resistance onto C3H mice through crossing and backcrossing to "carry along" the gene of heat resistance as well?

4. Is it possible to describe the extent of the damage to the heat center in the hypothalamus in biochemical terms? In other words, are substances formed in the brains of mice as a result of virus infection? Is the level of these substances, particularly neurogenic monoamines, increased or decreased in C3HRv as compared to C3H mice?

The list of puzzles is by no means complete. Whether any solutions will be forthcoming depends on whether we shall follow Butler's advice: "There is nothing so imprudent or so improvident as over-prudence or over-providence."

BY WAY OF AN ENVOI

Quand j'étais jeune, on me disait: Vous verrez quand vous aurez cinquante ans. J'ai cinquante ans; je n'ai rien vu.[6]   ERIK SATIE

The Copenhagen-Paris airplane, flight 1947, has been hijacked to 1969 by the worst pirate—time. André paints, I play the piano. Do we still have the *choix des élus,* or is the plane flying against the winds, carrying us nobody knows where? For seriousness, the past is more palatable than the future. But there is still the laughter which rings from every audience André had. His performance would rate an entertainment tax, had this problem not been solved by the decision of Mr. Justice Wool in *H. M. Customs and Excise* vs. *Bathbourne Literary Society.* Here is his momentous decision:

It [belief in the illegality of laughter] springs from two wrongful but widely spread beliefs: first, that what is instructive cannot be amusing; and, second, that what is amusing ought not to be allowed and, if it cannot be prevented, ought to be discouraged.

The first is the common fallacy of confusing heaviness with weight, of supposing that a light touch is the same thing as levity. Statesmen who make the House of Commons laugh are almost always suspected of insincerity and shallowness. A speech which is witty we are inclined to dismiss as so much

[6] Quoted in: Ned Rorem, *The Paris Diary of Ned Rorem;* with a portrait of the diarist by Robert Phelps, New York, George Braziller, 1966, p. 136.

"froth," forgetting that froth is the best sign that there is good beer below. But a dull speaker, like a plain woman, is credited with all the virtues, for we charitably suppose that a surface so unattractive must be compensated by interior blessings. . . . Judgment for the defendant.[7]

REFERENCES

Goodman, G. T., and H. Koprowski (1962). *J. Cell. Comp. Physiol.*, **59** (3), 333.
Gröschel, D., and H. Koprowski (1965). *Arch. Ges. Virusforsch.*, **18**, 379.
Hanson, B., and H. Koprowski (1969). *Microbios*, **1B**, 51.
Sabin, A. B. (1952). *Ann. N.Y. Acad. Sci.*, **54**, 936.
Webster, L. T. (1937). *J. Exp. Med.*, **65**, 261.

[7] A. P. Herbert, *Uncommon Law*, London, Methuen & Co., 1936, p. 412.

L'évolution physiologique:
a retrospective appreciation

R. Y. STANIER

In *L'évolution physiologique: étude des pertes de fonctions chez les microorganismes* (1944), André Lwoff summed up the knowledge and insights acquired during two decades of scientific work. It marks the end of his studies on protozoa and bacteria. Written during the occupation of France, and hence inaccessible to the outside scientific world for several years, the book was something of a war casualty. About 1950, I tried to persuade the author that a revised English version would be very useful, but he was far too absorbed in exploring the mysteries of lysogeny to entertain the suggestion. *L'évolution physiologique* greatly influenced my own scientific thinking, and so I should like to pay it the belated honor of a retrospective review.

As E. M. Forster (1927) has pointed out, "Books have to be read (worse luck, for it takes a long time): it is the only way of discovering what they contain. A few savage tribes eat them, but reading is the only method of assimilation revealed to the West." Heedful of this sage advice, I have reread *L'évolution physiologique* with pleasure and profit. I fear, however, that it could not be safely recommended to anyone under the age of 35, for whom biology begins with the double helix (Watson and Crick, 1953). Sad as it may be for the survivors, revolutions have a way of drawing a veil over the past; it is difficult for a postrevolutionary generation to understand the climate of the prerevolutionary period. Our science has changed so profoundly that the conceptual framework of thirty years ago must appear at once obscure and naive to the younger generation, if by chance they ever try to read the literature of that bygone era. Accordingly, my retrospective review will not be strictly factual, but will rather take the form of a historical meditation on some of the themes exposed in *L'évolution physiologique*.

The book has many facets. In part, it is an elegant epilogue to a major phase in the growth of classical biochemistry: the identification of vitamins and the establishment of their roles in cellular function, a development that roughly spanned the two decades between the first and the second world wars. Although most vitamins were discovered as essential nutrients for animals, the gross symptoms produced by their deficiencies provided only the faintest clues to their metabolic roles. The recognition that vitamins are also indispensable nutrients for many microorganisms was an essential precondition for the analysis of their functions. André and Marguerite Lwoff played a leading part in determining the vitamin requirements of microorganisms, largely (but not exclusively) in the context of the protozoa, while the contemporary British school led by Fildes was doing the same job for bacteria.

From this, the Lwoffs went on to analyze the specific effects of vitamin deficiencies on microbial cellular metabolism. A key study on the requirement for hematin revealed its involvement in the oxygen-linked respiration of trypanosomes (M. Lwoff, 1940). Five years previously, Warburg, Christian, and Griese (1935) had determined the chemical structure and mode of action of TPN, which immediately suggested that the vitamin nicotinic acid served as a biosynthetic precursor of pyridine nucleotide coenzymes. Shortly afterwards, the Lwoffs (A. Lwoff and M. Lwoff, 1937) showed that the so-called V-factor requirement of *Hemophilus* species could be met by these coenzymes; this was the first demonstration that an intact coenzyme may function as a vitamin.

André and Marguerite Lwoff accordingly provided many of the essential links in the chain of evidence which eventually led to the recognition of the relationships between vitamins and coenzymes, and hence of the role of vitamins in cellular metabolism. These discoveries immediately permitted another important conclusion: organisms that do not require a given vitamin must be able to synthesize it. Lwoff was one of the first to appreciate this fact and its wider implication: namely, that vitamins constitute a subclass of the larger class of *organic growth fac-*

tors, defined as essential biosynthetic building blocks which an organism cannot synthesize *de novo.* This extension of the original vitamin concept is expressed with exemplary clarity in *L'évolution physiologique:*

Les quelques exemples étudiés permettent de dégager une vue d'ensemble des facteurs de croissance.

Les facteurs de croissance sont des métabolites essentiels au sens de Fildes, c'est-à-dire des substances en l'absence desquelles la vie est impossible. La plupart d'entre eux sont des coenzymes ou des parties constitutives de coenzymes. Ils jouent à ce titre un rôle fondamental dans un ou plusieurs processus indispensables. La vie implique la mise en jeu d'un très grand nombre de réactions. Qu'une seule ne puisse avoir lieu et toute la machine se trouve arrêtée. Si, par exemple, l'un des rouages nécessaires au transport de l'hydrogène vient à manquer, le métabolisme sera entièrement inhibé. Si les coenzymes sont synthétisés par la cellule, un apport exogène n'est pas nécessaire. Si cette synthèse n'est pas réalisée, le coenzyme en cause devient un facteur de croissance. Ceux-ci rappelons-le ont été définis: des substances indispensables en l'absence desquelles toute multiplication est impossible et dont l'organisme est incapable de réaliser la synthèse.

Mais les facteurs de croissance ne sont pas forcément des coenzymes. Considérons par exemple le cas d'un acide aminé. Le tryptophane fait partie de la molécule des protides et prend par conséquent part à la construction de la molécule des enzymes. Il est logique de considérer que le tryptophane est un facteur de croissance pour un organisme n'en réalisant pas la synthèse. Il n'y a aucune raison de séparer les constituants des apoenzymes de ceux des coenzymes et il vaut mieux donner des facteurs de croissance une définition large qui fasse appel uniquement à leur nature de métabolite essentiel et au fait qu'ils ne sont pas synthétisés par l'organisme.

Les facteurs de croissance sont donc des métabolites essentiels que l'organisme est incapable de synthétiser.

Although Lwoff did not then know it, by 1944 the next major extension of these discoveries had already been made. Beadle and Tatum (1941) had embarked on the systematic isolation of induced auxotrophic mutants in *Neurospora,* thus opening one of the routes to our present knowledge of gene function, and at the same time providing a powerful tool for mapping the primary pathways of biosynthesis. It should be noted that, within the rather severe limits set by the biological and chemical materials then available, the Lwoffs had already begun to explore the biosynthesis of vitamins, in particular through their analysis of the specific components of the thiamine molecule necessary to meet the natural requirements of various microorganisms for this vitamin.

From what I have so far written, the younger generation might suspect that in prewar years André Lwoff belonged to that now discredited tribe, the classical biochemists. This inference would be incorrect: his biochemical knowledge was secondarily acquired, in part through sojourns at Heidelberg with Otto Meyerhof and at Cambridge with David Keilin. The truth is, if possible, worse: he was a member by formation of an even more discredited tribe, the classical zoologists. Thanks to his early association with Édouard Chatton, to whom the book we are considering was dedicated, he possessed an extraordinary knowledge of the protozoa as a biological group. This background enabled him, as almost none of his biochemical contemporaries could have done, to integrate the emerging knowledge of growth factors in the framework of general biological theory. He perceived (Lwoff, 1932) that microbial growth factor requirements can often be plausibly interpreted as an evolutionary consequence of adaptation to a parasitic mode of existence. They are one manifestation of a regressive physiological evolution, the counterpart in terms of function of the regressive structural evolution long recognized by zoologists as marking the development of parasitism in protozoa and many invertebrate phyla. Such an idea must have come easily to a student of Chatton, whose studies on parasitic dinoflagellates had furnished a beautiful illustration of regressive structural evolution associated with parasitism. This important theme is developed *in extenso* in *L'évolution physiologique,* which is still one of the best general treatments of the subject.

Among bacteria, which do not have much to lose in the way of structure, regressive evolution in the context of parasitism is usually expressed in physiological terms, as described by B. C. J. G. Knight (1936) in another classic book of the same period. Some years ago, we discovered a particularly clear case of regressive evolution in the context of bacterial parasitism, which had both structural and nutritional manifestations (Redfearn, Palleroni, and Stanier, 1966). Since it illustrates one of the main themes of his book, I should like to offer a brief recapitulation of the facts to André Lwoff.

The few *Pseudomonas* species pathogenic for vertebrates include two, *P. pseudomallei* and *P. mallei,* which we showed to be remarkably similar in phenotype, and which have subsequently been found to be also closely related genetically (Rogul et al., 1968). *Pseudomonas pseu-*

domallei is the causative agent of melioidosis, an almost invariably fatal and untreatable disease which occurs only in regions of the earth lying between 20°N and 20°S. Consideration of its nutritional phenotype, which we were the first to determine, as well as a review of the epidemiology of the exotic disease for which it is responsible, convinced us that *P. pseudomallei* is not a parasite: it belongs to the category of accidental pathogens, not uncommon among the bacteria. This species is almost certainly a normal member of the tropical microflora of water and soil but can occasionally establish traumatic infection in animal hosts, particularly if their level of natural resistance is low.[1] Since accidental pathogens are almost never transmitted from host to host, they are subject to natural selection only in the context of their normal habitat.

Pseudomonas mallei, the agent of glanders, is on the other hand a strict parasite of its equine hosts. The phenotypic differences between these two species revealed by our study are represented almost exclusively by characters that are possessed by *P. pseudomallei,* but lacking in *P. mallei.* The parasitic member of the pair has lost the ability to use many of the varied organic carbon sources utilized by its nutritionally versatile congenor. It is also unable to synthesize flagella, being the only known *Pseudomonas* species that is permanently immotile, a structural regression that had prevented recognition of its correct taxonomic position prior to our work.

In the final chapters of *L'évolution physiologique,* Lwoff attempted to extend the concept of regressive physiological evolution beyond the context of parasitism and to apply it to biological evolution in the widest sense. This involved him in a somewhat risky comparison between biological and physical evolution, in which physiological regression was analogized with the increase of entropy. Implicit in this treatment was the notion that early biological systems were endowed with a high level of synthetic ability, subsequently whittled away by irreversible losses of function. In many specific evolutionary contexts, the notion is extremely useful: for example, in the interpretation of the evolutionary origin of protozoa from photosynthetic ancestors through the loss of chloroplasts,

---

[1] An obvious prediction (which we did not mention in our article) was that melioidosis would become a significant military medical hazard as a result of the intensification of the war in Vietnam. This prediction has, unfortunately, proved correct, although the fact has not been publicized by American authorities.

a subject that is beautifully discussed in the book. As a general concept, however, regressive physiological evolution contains a formidable paradox: the assumption of maximal metabolic complexity at the outset of biological evolution. I cannot help feeling that Lwoff's treatment of this paradox was evasive: "il est vain d'épiloguer sur un passé inaccessible à nos moyens d'investigation." Since this is precisely the complaint of our more sceptical colleagues, for whom any evolutionary speculation has a faintly scandalous, *fin de siècle* quality about it, this conclusion was not very fitting for a book replete from beginning to end with evolutionary speculations.

Today, however, we can appreciate that Lwoff's general view of the course of physiological evolution represented only half the story. An essential complement, which resolved the paradox of Lwoff, was the later hypothesis of Horowitz (1945) concerning the way in which biosynthetic capabilities of ever-increasing complexity could have been acquired by living systems in the course of primary biochemical evolution. The generally negative slope of physiological evolution perceived in 1944 by Lwoff now assumes its proper place as one component of a more complex evolutionary curve, which had a strongly positive initial slope.

I cannot resist ending this essay with the quotation of a delightful and typically Lwoffian passage from the introduction to his book. It might be entitled "The Gentle Art of Making Enemies" or, perhaps, *On ne badine pas avec l'amour-propre*.

La doctrine de l'évolution physiologique régressive, exposée pour la première fois en 1932, fut en général mal accueillie en France. Non qu'elle ait été ouvertement combattue: à quelques exceptions près, elle rencontra plutôt de l'indifférence. Nous nous aperçûmes, au cours de conversations avec maîtres, collègues et amis, zoologues ou botanistes, que l'idée d'une dégradation physiologique évolutive heurtait profondément les esprits. L'un d'eux nous fit comprendre que nous ignorions tout de l'évolution: nous ne partagions pas ses idées. Un autre nous demanda de reconnaître qu'il était susceptible de réaliser des actions dont un *Chlamydomonas* était totalement incapable. Sans doute. Cependant, que serait devenu cet excellent—et par ailleurs éminent—collègue si on l'avait exposé au soleil dans une solution de nitrate de potassium?

REFERENCES

Beadle, G. W., and E. L. Tatum (1941). *Proc. Nat. Acad. Sci.,* **27,** 499.

Forster, E. M. (1927). *Aspects of the Novel,* New York, Harcourt and Brace.

Horowitz, N. (1945). *Proc. Nat. Acad. Sci.,* **31,** 153.

Knight, B. C. J. G. (1936). *Bacterial Nutrition,* Med. Res. Council (Brit.) Spec. Rep. Series No. 210, London, His Majesty's Stationery Office.

Lwoff, A. (1932). *Recherches biochimiques sur la nutrition des protozoaires,* Paris, Masson.

Lwoff, A. (1944). *L'évolution physiologique: étude des pertes de fonctions chez les microorganismes,* Paris, Hermann.

Lwoff, A., and M. Lwoff (1937). *Proc. Roy. Soc. (London),* Ser. B, **122,** 352.

Lwoff, M. (1940). *Recherches sur le pouvoir de synthèse des flagellés trypano-somides,* Paris, Masson.

Redfearn, M. S., N. J. Palleroni, and R. Y. Stanier (1966). *J. Gen. Microbiol.,* **43,** 159.

Rogul, M., J. Brendle, D. K. Haapala, and A. D. Alexander (1968). *Bact. Proc.,* p. 19.

Warburg, O., W. Christian, and A. Griese (1935). *Biochem. J.,* **282,** 157.

Watson, J. D., and F. H. C. Crick (1953). *Nature,* **171,** 737.

Back to pangenesis?

M. R. POLLOCK

"Viruses are viruses."   A. LWOFF, 1957

Unfortunately for me, but fortunately, I think, for him, I never had the advantage of working directly with André Lwoff. I was, however, a visiting worker in his department at the Institut Pasteur on two occasions and came, under the influence of his charm and gentle (*usually* gentle) wisdom, to develop a deep respect and affection for him, both as a human being with a delightful sense of humour and as a great biologist.

So I feel rather privileged to have been invited to contribute to this collection of essays in his honour. And, quite illogically, I propose to interpret this privilege as permission to indulge in a more personal, rambling and philosophical article than I might feel was required of me had I ever been a research collaborator like most of the other authors in this volume.

My epigraph is a quotation because André likes quotations (Lwoff, 1953); a quotation from Lwoff because that is appropriate; a quotation that is a revelation, a meaningless tautology, a profound truth, etc., according to one's mood or philosophy; a quotation that André himself, as much as anyone else, has shown, I would maintain (at great personal risk), to be untrue—untrue, that is, in the Baconian sense. After all, with the discovery of prophage, a child could answer the conundrum, When is a virus not a virus?

The first time I met André was shortly after World War II, in the excitement of renewed contact with Europe after more than 7 years of isolation. This was at the beginning of the Golden Period in the Service de Physiologie Microbienne at the Institut Pasteur in Paris: the 10 Golden Years in those cramped laboratories (before Jacques Monod and his *équipe* abandoned the top floor for more spacious apartments downstairs) with close, excited contact between so many stimulating and egregious characters. This was the time when "enzyme adaptation" was still *soft,* molecular biology not yet invented, or induction, in any form apart from embryonic or philosophical, even contemplated: the heroic decade before Watson-Crickery had cramped the freedom of our imaginations.

The minor actors in the drama of those inspiring and perplexing days may be forgiven for feeling some nostalgia for the time when they still believed that they themselves might conceivably make some small contribution to the sum of human knowledge: before repressors and operators had accumulated for the purpose of telling our genes what to do with themselves; before the Code was invented, let alone solved; before permeases, plasmids, prophages, and promoters had swamped our minds to scour our ideas and decontaminate our thoughts from any germs of nonmolecular heterodoxy.

It was during my first stay at the Pasteur, early in 1948, that I began to learn a little of the delights—and the dangers—of working in that famous laboratory. The pressure was heavy (or so it seemed to me) to appear brightly intelligent and quick-witted enough to parry the cut and thrust of these terribly clever French scientists who so enjoyed teasing their rather amusing Anglo-Saxon country cousins. One had to choose weapons and tactics carefully, and woe betide any attempt to make fun of the French (as distinct from their governments, which were legitimate targets for withering sarcasm, even from foreigners) in the way the French made fun of the English. But there was so little malice in this bantering that it was an educative experience in itself and a small price to pay for all the kindness and consideration shown to the visitors and for the stimulating conversation that was daily fare at the informal laboratory lunches. These were often presided over by Jacques Monod, whose opinions—on absolutely *everything*—were always listened to with great respect and interest, were always worth hearing, and indeed could hardly avoid being heard.

Things were often very difficult for French research workers at that time. Little equipment was available (most of the worst gaps, however, being bridged by grants from the United States), and salaries were low. Conversation often took the form of criticism of the immovable French Academic Establishment, which did not seem to accept this particular group of Pasteur scientists as proper representatives of *la civilisation française*. This may have led to unnecessary frustrations, but I suspect that the workers concerned would have been rather disappointed had they been so accepted and thus deprived of a legitimate target for their witticisms.

In those days Jacques was the only French worker in the laboratory who spoke perfect English; that is, unless one includes Louis Rapkine, who often used to join us for lunch. What a great pleasure it was to talk to someone so generous and philosophically charming as Louis seemed to me to be. And this was only a few months before his tragically premature death.

Jacques was always liberal with his English. With André, however, French was *de rigueur,* though even in those early days he could, I suspect, speak English better than I could French. When, in conversation, André broke into English it seemed a terrible disgrace because the reason was clearly that I was failing—failing disastrously—in the language of the country in which I was working. But nobody could have been more personally generous and considerate, if a trifle impatient, than André in helping our family (with three and a half children) through all the difficulties of life in Paris shortly after the war. Not the least of these was the French rationing system, a rather disorganized and quite unpredictable affair. Transport was by Metro (which always seemed either to be impossibly overcrowded because of bus strikes, or to be on strike itself). Trips outside the city were almost invariably by Ligne de Sceaux out to St. Rémy de Chevreuse, where we usually managed to drag the children half a mile or so from the station before collapsing for a picnic lunch by the roadside.

But, of course, it was all absolutely marvellous: living in Paris (Montmartre) at a time when *no* foreign currency was normally available for Britons (or petrol in the United Kingdom), and working at the fabulous Pasteur in the company of the famous Lwoff (one could mention Jacques Monod, with whom, of course, I was actually working, but his name seldom produced much effect in those days), and still being young enough to feel one might one day achieve something important. Lack of independent transport meant less than it would have done now, and André would occasionally take our whole family out in his car (his was one of the very few amongst Pasteur workers) for a longer trip outside Paris. The flood of American visitors to the Institut had not then begun, of course, and I was very privileged.

Life was much easier during my second visit in 1952–1953, and work at the laboratory more prolific and scientifically more exciting. Mel Cohn was almost in command, the corridors ringing with his laughter and atrocious French accent.

Those were the days of the Terrifying Seminars. Unsuspecting visitors would be trapped by charm and flattery into revealing their imperfections by giving a seminar before a rather devastating audience in André's office. I remember one innocent Anglo-Saxon, passing through Paris on a *holiday,* poor fellow, being lured into such a performance. Knowing all too well the nature of these affairs, the subject chosen (enzyme adaptation itself, no less), and the relatively unsophisticated temperament of the victim, I tried to warn the unsuspecting visitor what he was in for, but to no avail. The prospect of this great opportunity was too much for him. He plunged in headlong, barely bothering to marshal his ideas or collect his facts. Needless to say, he emerged scarcely breathing, with his theories hanging from his bruised body in tattered shreds.

They were tough going, those seminars. Though sometimes rather cruel, they were, however, exhilarating and stimulating and, I believe, on the whole justified. It just was not possible to "get away" with something half-baked or imprecise, and we all learned a great deal from them.

In fact, they were probably more a creation of Jacques's (or at least a reflexion of his personality) than of André's. Superficially, André seemed to be more in the background and to exercise a somewhat restraining effect, and it was not until later that I fully realised the tremendous extent of his influence in that fantastic laboratory, which originally was almost entirely his own creation. As a nidus of sheer brilliance it can hardly ever have been surpassed.

All this was still really before the great breakthrough. Although everything was becoming more and more molecular, "molecular biology" was still just a rather useful name for the preoccupation of a small group of workers at the Cavendish Laboratory in Cambridge, which few of us had heard of.

For André, it was never really necessary to become "molecular," however much (as he himself has emphasised) it may since have become fashionable. His approach to a problem has often seemed brilliantly intuitive rather than consciously hypothetico-deductive. And, for me, his attitude to his own work has always appeared to be in the classical tradition of biology. He often analyses his immensely original discoveries as contributions to "virology" and the sharpening of its definition, using terms, such as "parasitic" and "pathogenic," which for many have become almost meaningless within the framework of the particular system concerned. This, from a personal point of view, is an interesting and perhaps significant fact; and of course it detracts nothing from the importance of his researches, the fascination of which, to many of us, has been the new vista of intra- and intercell interactions thus disclosed. For many molecular biologists the discovery of lytic phage induction in lysogenic bacteria, the characterization of the lysogenization process, the immunity relationships in phage infection, the concept of prophage itself, and all the exciting recent developments in λ-prophage genetics and biochemistry, which stem directly from André's original work, have been major factors leading to a new orientation of outlook on cell physiology.

This modern bacteriophage biology and much of the recent work in the general biochemical dissection of the cell have inevitably led to a reappreciation of the immensely delicate and complex coordination between the various intracellular elements that must be necessary to ensure integrative function of the whole system. At the same time, apparently paradoxically, the extent of relative autonomy in the function and replication of separate nucleic-acid-containing elements within and between cells has emerged in a fashion so striking that in many respects the cell itself has come to be viewed more as a mosaic of interrelated, but distinct and potentially independent, entities than as a tightly integrated whole entirely dominated by a central nuclear authority (Pollock, 1969).

More significant still, however, is the increasing realization of the extent of nucleic acid interactions between different types of cells. And by this I do not mean the intraspecific exchange and recombination of genetic information involved in sexual reproduction, which involves relatively slight modification to the genome. I am thinking rather of events of a wider significance, caused by interspecific transfer of genetic material and gain or loss of, or large quantitative alterations in, relatively independent elements such as episomes, plasmids, and other DNA-containing cytoplasmic organelles (mitochondria, chloroplasts, etc.).

The fact that such heterologous transfers are relatively rare only emphasises their significance. Natural barriers to the "intrusion of foreign nucleic acids" have, of course, evolved for the simple teleological reason that most such transfers would be metabolically disastrous. But they must now be regarded as a great deal commoner than was originally realised. "Virus infections" are not necessarily (or even usually) pathogenic. Interspecific transfer of plasmids is relatively easy, though it seems sometimes to involve weakened control of plasmid numbers per cell (Rownd, 1969) and leads to large increases in plasmid gene copies, which could be of great significance. Duplication of specific chromosomal genes can now be forced by suitable selection techniques (Beckwith, Signer, and Epstein, 1966). Interspecific hybrid cells can be constructed amongst vertebrates by relatively simple manoeuvres for overcoming the nucleic acid intrusion barriers (Ephrussi and Weiss, 1965; Harris, 1966) and can then lead to heterologous physiological repair of chromosomal markers, if not specific recombination events (e.g., between chromosomes of mouse and man: Weiss and Green, 1967). Even the process of antibody formation has been alleged to involve specific nucleic acid transfer between two highly differentiated types of cell (see Fishman, 1969). And the frequency of so-called helper viruses necessary for the normal intracellular growth of many plant and animal viruses (see Kassanis, 1968) emphasises this mosaic quality amongst and between viral genomes themselves.

Our concept of the organism and of the significance of the cell as an integrated unit in the biosphere is undergoing a revolution, and it is largely the result of what we have come to know about the behaviour of isolated pieces of nucleic acid. Amongst these must be included most of the elements now referred to as viruses.

But there is really little new in this idea of a "mosaical" cell.

Approximately 100 years ago there occurred two interesting events of some relevance to this subject. I am not referring, as might at first be expected, to the famous paper in 1866 by Mendel on pea genetics, or to the publication in 1859 of the *Origin of Species*. Rather, I am citing another work by Charles Darwin (1868), having the title *Variation of Animals and Plants under Domestication* (Vol. II), in which he developed his theory of "pangenesis," and also the article by Louis Pasteur (1861) wherein he finally disposed of the once dominant belief in spontaneous generation.

One might argue, of course, that induction of phage in a lysogenic bacterium was a *form* of spontaneous generation, but it is hardly of the type that Pasteur was dealing with. We take for granted that, if we run into a biologically active piece of nucleic acid (be it in the form of virus, gene, mitochondrion, etc.), it must have been genetically derived, with only minor modification, if any, from a previously existing biosystem. In that sense, Pasteur's work is almost the axiomatic foundation of all molecular biology. (How shattered we should all be were some clay surface to be discovered with the property of templating the condensation of nucleotides in a biologically significant sequence!) Darwin's "pangenesis," however, means very little to anybody now, and few scientists would even be able to describe it. Yet, in essence, and with only one (admittedly vital) difference, it expresses many of the ideas gaining in current popularity at the moment, and it is exceptionally interesting to try to trace their origin and development.

"Physiologists agree," wrote Darwin (1868), citing both Claude Bernard and Virchow, "that the whole organism consists of a multitude of elemental parts which are to a great extent independent of each other." Elsewhere in the same work he states, "Each living creature must be looked at as a microcosm—a little universe, formed of a host of self-propagating organisms, inconceivably minute and as numerous as the stars in heaven." It is quite obvious, from the context, that the "organisms" of which he speaks are not cells but the "gemmules" postulated in his theory of pangenesis. These gemmules correspond, in some ways, to genes but differ in two important respects: they could undergo heritable modifications under the influence of the environment and they could pass fairly freely from cell to cell, finally aggregating in the gametes. Ironically enough, Darwin was almost obsessive in his Lamarckian belief in the inheritance of acquired characters, and in any case those were the days before Weismann's theory of the separate germplasm came to dominate biology.

It was actually the more realistic and more reasonable modification of Darwin's gemmule hypothesis by de Vries (1887), in the form of what he called "intracellular pangenesis," that brought these ideas closer to modern ways of thinking. De Vries refused to accept the idea of gemmules (which he renamed "pangens") passing from cell to cell, and he elaborated their properties in a manner which brings them closer to modern genes, differing mainly in his overemphasis of their reproductive and functional independence. However, he dealt at some length with what he called "compound" gemmules—groups of particles, members of which "become active at the same time" (shades of the operon?).

It is relatively easy, of course, and often very misleading, to draw close analogies between ideas of the past and modern theories of the present. One could argue that present concepts in molecular genetics have no more connection with Darwin's and de Vries's ideas of "pangenesis" than modern atomic theory has with Democritus' atoms.

But the recurring cycle of emphasis on significant levels of biological organization is a striking one. More than 80 years ago de Vries (1887) was accepting the relative autonomy of most intracellular organelles, such as "plasmatophores" (plastids?), "amyloplasts," vacuoles, and ("probably") the cytoplasmic membrane, on what was clearly very insufficient evidence. Yet it was only 5 years or so ago that the independent replicative powers of chloroplasts and mitochondria were fully accepted again.

Perhaps, after all, there may still be something left to explore and renovate in Darwin's pangenesis idea, whether we designate his gemmules as viruses, episomes, plasmids, or genes. There is, at all events, much more intercellular transfer and reassociation of bits of nucleic acid (be they wrapped up neatly in protective protein coatings, wriggling painfully through pili, or floating freely as naked and nubile nucleotides in their wanderings from cell to cell) than we had previously dreamed of.

"I feel sure," wrote Darwin in his letter to J. D. Hooker in 1868, "that if pangenesis is now still-born it will, thank God, at some future time re-appear, begotten by some other father, and christened by some other name." Not for one moment would I suggest that we have yet reached the stage of renaming the theory of pangenesis or of recognising clearly its modern counterpart as some aspect of virology, let alone identifying its father. But maybe we are developing a little way in that direction.

It becomes understandable why the individuality of viruses, and their uniqueness, have been so strongly emphasised (especially with reference to phage) by André (in his insistence that "viruses are viruses") after reading some of the earlier papers on the "Twort-D'Herelle" phenomenon by such people as Bordet, Twort himself, and Eugène and Elizabeth Wollman. Many of the conceptual difficulties involved seemed then to arise from attempts to classify the phage phenomena that we now, thanks largely to André himself, understand so well, *simply* as manifestations of bacterial genes *or* of bacterial parasites, that is, as types of processes already familiar and well known.

When certain groups of events begin shaping themselves persistently and consistently into a new pattern, it is usually time to recognise this explicitly by admitting a new name and elevating the pattern into a new concept. Although originally viruses were regarded simply as intracellular parasites rather smaller than bacteria, the sophistication of certain aspects of virus-associated phenomena, such as viral immunity, lysogeny, latent virus, phage conversion, and viral transformations, indicates that a new type of pattern may indeed be waiting to be defined and recognised.

The Third Marjory Stephenson Memorial Lecture (Lwoff, 1957) was the first attempt to deal with these problems. It was based on a modernized version of the concept of virus, carrying with it the original implication of a distinct and separate entity which was, however, variable in expression, moving around the place in several different states and disguises.

One reason, perhaps, why this idea is not entirely satisfying to some persons is that the expressed behaviour and physical state of this entity depend absolutely (directly and indirectly) upon the nucleic acid environment with which it is associated (cooperatively or antagonistically) at any given moment. For instance, the main λ-phage-specific DNA sequences can function to a certain extent and be replicated as a chromosomal gene complex (prophage), an agent facilitating genetic recombination between two types of bacterial cell (in transduction), a lytic cytocidal virus, or (as recently demonstrated by Signer, 1969) an extrachromosomal plasmid, replicating coordinately with the bacterial cell.

These phenomena are regarded as facets of viral behaviour. The concept is equivalent to the currently accepted notion that a gene is simply a gene, whether it be integrated with thousands of others in a chromosome, undergoing autonomous replication and transfer as part of a plasmid or F factor, wrapped up in a phage head, or floating freely in the cytoplasm or in the extracellular environment. To some, however, the biological significance of a gene may be fundamentally changed by the nature of its immediate associations, especially by the extent to which it may, or may not, be capable of replication and in what context.

Burnet's (1957) observation that "a virus is . . . something which could almost be called a stream of biological patterns" is quoted by André Lwoff (1957) with approval at the end of the Memorial Lecture mentioned above. I believe many would feel that this is the right *sort* of approach, but far too vague and amorphous to do anything more than set the direction of our thinking. I would prefer something indicating a *specific focus of nucleic acid interaction*.

"Viruses may be viruses" from a strictly logical and etymological point of view, and it is surely right to avoid thinking about them as "escaped genes," etc. We seem, however, to be moving into a conceptual world where the areas of most *significant* behaviour are those in which nucleic acid entities can interact and replicate with metabolically coordinated results for long periods of time (as indeed emphasised later by Lwoff, 1960, himself), whatever their origin and whatever the form in which they meet. So we still need to develop our ideas and definitions further.

But if it is really essential to choose a material entity rather than a "focus of interaction" as a new centre of attention, an alternative approach might be to retrace our steps and simply consider the possibility of concentrating on the biology of *small replicons*. The evolutionary history and physiological behaviour of such elements must reflect the need for many large, preformed building blocks in the immediate environment if replication is to proceed. This requirement, which is simply a reflection of the limited amount of information available in the replicon genome, would lead to the evolutionary development (*a*) of special mechanisms for the exploitation and diversion of energy, building blocks, and enzymes provided by other biosystems containing the necessary information, and (*b*) of sophisticated apparatus to facilitate the passage of the replicon from one biosystem to another (perhaps more suitable). There are, of course, objections to the definition of a "small replicon" being fixed arbitrarily purely on size. But it might perhaps be determined by the limit at which some specifically significant biochemical reaction (e.g., the high-energy-yielding, so-called Lipmann system suggested by André Lwoff, 1957, in a rather different context, or the ribosome-tRNA complex of Lwoff and Tournier, 1966) was available only through information provided by some other biosystem.

One effect of this reorientation, of course, is that certain phases of the viral life cycle as at present defined would be excluded, and a host of plasmids and episomes brought in. How outrageous! However, I hope André will forgive me, even if he thinks this is all nonsense, and bring himself to accept it as part of an attempt, crude and primitive though it be, to pay tribute to the many great things he has done for the science of (molecular) biology.

REFERENCES

Beckwith, J. R., E. R. Signer, and W. Epstein (1966). *Cold Spring Harbor Symp. Quant. Biol.*, **31**, 393.

Burnet, F. M. (1957). Sci. Amer., **196**, 37.

Darwin, C. (1868). *The Variation of Animals and Plants under Domestication*, vol. II, Chap. 27, London, John Murray.

Ephrussi, B., and M. C. Weiss (1965). *Proc. Nat. Acad. Sci. U.S.*, **53**, 1040.

Fishman, M. (1969). *Ann. Rev. Microbiol.*, **23**, 199.

Harris, H. (1966). *Proc. Roy. Soc. (London)*, Ser. B, **166**, 358.

Kassanis, B. (1968). *Adv. Virus Res.*, **13**, 147.

Lwoff, A. (1953). *Bact. Rev.*, **17**, 270.

Lwoff, A. (1957). The Third Marjory Stephenson Memorial Lecture, *J. Gen. Microbiol.*, **17**, 239.

Lwoff, A. (1960). The Leeuwenhoek Lecture, *Proc. Roy. Soc. (London)*, Ser. B, **154**, 1.

Lwoff, A., and P. Tournier (1966). *Ann. Rev. Microbiol.*, **20**, 45.

Pasteur, L. (1861). *Ann. sci. nat.*, **16**, 5.

Pollock, M. R. (1969). The changing concept of organism in microbiology, *Progr. Biophys. Mol. Biol.*, 19, 273.

Rownd, R. (1969). *J. Mol. Biol.*, 44, 387.

Signer, E. R. (1969). *Nature*, 223, 158.

Vries, H. de (1887). *Intracellular Pangenesis*, translated by C. Stuart Jager, Chicago, The Open Court Publishing Co., 1910.

Weiss, M. C., and H. Green (1967). *Proc. Nat. Acad. Sci. U.S.*, 58, 1104.

The unmasking of the unseen

NIELS OLE KJELDGAARD

Only that day dawns to which we are awake.

HENRY DAVID THOREAU

Studies of lysogenic bacteria have had a long tradition at the Institut Pasteur. After the end of World War II, André Lwoff resumed the fundamental work on the lysogenic *Bacillus megatherium* of Elisabeth and Eugène Wollman (1936, 1937, 1938), which was so tragically interrupted in 1943. It was my privilege in 1949–1950 to participate in one of the very successful phases of this work: the induction of bacteriophage production and mass lysis of lysogenic bacterial cultures.

It was late September 1949 when I learned that I had won a fellowship from the French government to spend ten months at the Institut Pasteur. Two weeks later I was on my way, filled with the high hopes and joy of a young man who, after the confinement of the war, for the very first time left his home country, heading for the metropolis of revolutionary spirit, dreams, and action. To be honest, I have to admit that my desire to go to Paris was as much influenced by the walks of Jerphanion and Jallez as by the hope for scientific achievements.

After a few days of drowning myself in impressions of the city of Sainte Geneviève and Thorez I found my way to Rue du Dr. Roux and mounted the stairs to the attic of the research building of the Institut Pasteur. Having penetrated through the dark labyrinth of the laboratory of Dr. P. Grabar to the Service de Physiologie Microbienne, I was ushered in to Dr. Lwoff by his most kind secretary Gisèle, who as a friendly dragon guarded the entrance of his office against intruders.

That André Lwoff was working with the problems of lysogeny had not come to my knowledge before I left Denmark, and thus I arrived rather unprepared for the task to be accomplished. At that time I had been working for about two years as a graduate student in the laboratory of Dr. Herman Kalckar in Copenhagen; although the work of the laboratory was concentrated on studies of the enzymes involved in nucleotide metabolism, the marvels of the bacteriophages had made only little impact. Therefore, the introduction to lysogeny and to the work on microdrops which Lwoff gave me in the secluded quiet of his office on my very first day went, to a large extent, over my head. This situation changed during the first several weeks while I worked all by myself in the tiny laboratory next to the office to improve the growth media of the *B. megatherium,* learning the importance of exponential growth and getting acquainted with the people in the "Service Lwoff."

Compared to more recent times, the group of 1950 was pleasingly small. The team working on lysogeny included, besides *le patron* and his two technicians, Antoinette Gutmann and Évelyne Ritz, Louis Siminovitch and myself, both of us most wonderfully assisted by Mme. Sarah Rapkine. Further down the dirty-yellow corridor, the dynamic adaptive enzyme team was forcefully headed by Jacques Monod and included Anne-Marie Torriani and Melvin Cohn. For part of the year Michael Doudoroff added his flavor to the team. More lonely stars among the clusters were Pierre Schaeffer, Catherine Fowler, and Helen Ionesco.

Although its members were working with many different problems, the group was sufficiently small to allow a close contact between everyone and also to include outsiders in the family. Frequently François Gros ascended from Macheboeuf's laboratory on the first floor, Luigi Gorini came from Fromageot's lab at the Sorbonne, and George Cohen visited from the branch of Institut Pasteur in Garches. This openness to the scientific community seems to me one of the most remarkable aspects of the traditions of the Service de Physiologie Microbienne, which obviously has been of great importance to French biology.

At the end of the year 1949, the now classical work of Lwoff and Gutmann (1950) on the growth and phage production by single cells of the lysogenic *B. megatherium* had passed its most crucial phases. The extraordinary care and patience involved in this beautiful work hardly appears from their paper but became evident to me while I was learning the difficult art of depositing microdrops of medium in the oil chamber and inoculating the droplets with single bacteria. The last few experiments in the very extended series allowed me the opportunity to admire the dexterity and extreme skill of Dr. Lwoff. For whole days he would be seated in front of the thermostatted Plexiglas box housing the microscope and the micromanipulators. With a capillary pipette a few microns in diameter mounted in the De Fonbrune micromanipulator, newly divided bacteria were fished out of the droplets with a minimum of contaminating medium and transferred into fresh microdrops or deposited into a larger drop of medium to be handed to Antoinette Gutmann. By means of this cumbersome but direct approach, it was shown that lysogenic bacteria can multiply for numerous generations without free phages ever appearing in the medium. Under other conditions, one or several bacteria in the microdrops lysed and liberated a full burst of phages. Why this occasionally happened in some drops and not in others was a mystery. The stage was set for the hunt over the next several months.

For my own part the start of the hunt was rather slow, but early in 1950 we combined forces and I left my lonely laboratory to join Louis Siminovitch and Sarah Rapkine in one of the happiest collaborations of my scientific career. Under the general command of André Lwoff, we, as well as Antoinette Gutmann and Évelyne Ritz, tried to define the growth conditions which would allow mass lysis and phage production of the lysogenic *B. megatherium*. Growth curves were followed by optical density measurements in the rather primitive but useful Meunier photometer, and bacteriophage production during growth was determined after a thousand-fold dilution of the samples.

Almost every morning, as soon as all plaques were counted and the results plotted on graph paper, Louis Siminovitch and I went to see Lwoff in his office to discuss the score of the day and extend our list of experiments to be done. This daily discussion and digestion of the results normally concluded by Lwoff suggesting yet another model encompassing all available information. There were even days where the morning model was replaced by a late afternoon model.

Days were busy; like Siamese twins, Siminovitch and I would rush back and forth between the warm room and the lab, always juggling several experiments. The eventual discovery of mass lysis by ultraviolet irradiation did not come accidentally to the unprepared mind but was the result of systematic work and of a truly genuine and successful collaboration. In one of my protocols, prudently kept by Mme. Rapkine at the Institut Pasteur, one finds, besides my own handwriting, the hieroglyphs of Louis Siminovitch, the neat hand of Mme. Rapkine, and on most of the graphs calculations and comments by Lwoff.

Days were rushing by. I even made attempts to start my working day rather early, although I never beat the nine o'clock punctuality of Dr. Lwoff.

Days when inflections were observed on the growth curves and a rise in the phage titer was recorded brought high hopes. Such occasions were often celebrated by a drink of the rare aperitif "Banyuls," kept for the purpose in an Erlenmeyer flask in a cupboard in Lwoff's laboratory. *Le patron* did the pouring, a beaker for us, a trifle for himself.

Often, however, the attempted repetition of an experiment brought an end to our optimism. Early experiments suggested that the conditions of aeration were of importance for the phage production. This led to numerous experiments with measurements of redox potentials; variations in the media, involving especially the magic, homemade yeast extract; variations in the method of oxygenation; batch cultures of different volumes in Erlenmeyer flasks; cultures of different volumes in bottles; cultures with yeast extract, 20%, 10%, 3.5%, 2%, 1%; cultures with and without glucose; preutilized media; and bactogens aerated with air, air $+ CO_2$, nitrogen $+ 10\%$ air, nitrogen $+ CO_2$, everything in numerous combinations.

Eventually our list of experiments intended to influence the metabolism of the cells came to include irradiation with ultraviolet light. Our perseverance was fully rewarded when 60 seconds of irradiation from the spiral-shaped mercury lamp gave perfect lysis and phage production at the very first try. We all marvelled when Évelyne Ritz brought the news that the culture was lysing.

A whole sequence of experiments was an obvious and long-foreseen consequence. We knew that cells induced to produce phage and to lyse, somehow could recuperate. It was shown that starvation for an energy source was able to cure UV-induced cells. We suspected that not all conditions of growth made the cells likely to produce a burst of phages. The finding that only cells grown in a rich medium would lyse after UV irradiation clearly demonstrated the fundamental influence of the growth medium on the physiological state of the cells. This fundamental property of bacteria has almost earned me a living ever since (cf. Maaløe and Kjeldgaard, 1966). In fact, the publication of our studies on the lysogenic bacteria (Lwoff, Siminovitch, and Kjeldgaard, 1950) contains the results of one of the very first "upshift" experiments, in which the appearance of the aptitude in a minimal-medium culture was followed after enrichment of the growth medium by yeast extract and casamino acids. Since then, this type of experiment has been subject to detailed analysis (Kjeldgaard, Schaechter, and Maaløe, 1958).

> *There is more day to dawn.*   DAVID HENRY THOREAU

Twenty years have passed since the exciting induction days. The genetic background for the excision of prophage from the bacterial genome and the regulatory mechanisms governing phage production are rather well established, but curiously enough the primary target for the UV irradiation is still unknown.

My stay at the Institut Pasteur was all too short. Looking through the veil of memory, it might have been wise if I had accepted the invitation of Dr. Lwoff to stay for another period instead of leaving to finish my examinations.

If this short period is to constitute the background for summarizing the lesson taught me by André Lwoff, my account of it must contain an element of admiration for the noblesse of French culture and a reflection of the scientific attitude that models exist only to be discarded.

REFERENCES

Kjeldgaard, N. O., M. Schaechter, and O. Maaløe (1958). *J. Gen. Microbiol.,* 19, 607.
Lwoff, A., and A. Gutmann (1950). *Ann. Inst. Pasteur,* 76, 711.
Lwoff, A., L. Siminovitch, and N. Kjeldgaard (1950). *Ann. Inst. Pasteur,* 79, 815.
Maaløe, O., and N. O. Kjeldgaard (1966). *Control of Macromolecular Synthesis,* New York, W. A. Benjamin.
Wollman, E., and E. Wollman (1936). *Compt. rend. Soc. Biol.,* 122, 190.
Wollman, E., and E. Wollman (1937). *Compt. rend. Soc. Biol.,* 124, 931.
Wollman, E., and E. Wollman (1938). *Ann. Inst. Pasteur,* 60, 13.

Découverte d'une perméase dans un grenier

❧

GEORGES COHEN

J'avais environ vingt-cinq ans quand j'ai connu André Lwoff. Jacques Monod m'avait introduit au Club de Physiologie Cellulaire (qui devait adopter par la suite le nom plus rutilant de Club de Biologie Moléculaire). Après les réunions mensuelles, où nous écoutions et discutions le travail d'un conférencier dans la Bibliothèque de l'Institut de Biologie Physico-Chimique, nous allions dîner à la Brasserie Alsacienne du Boulevard Saint-Michel. Les dîners réunissaient souvent plus de vingt personnes autour d'une table du premier étage. André et Marguerite Lwoff étaient des commensaux de ces repas où l'on pouvait également rencontrer Jacques Monod, Élie Wollman, Louis Rapkine, Boris Ephrussi, Raymond Latarjet, Pierre Schaeffer, Germaine Cohen-Bazire, Piotr Slonimski, Alain Bussard, Luigi Gorini et des chercheurs étrangers venus passer une ou plusieurs années de stage à Paris: Seymour Cohen, Kits von Heyningen, Martin Pollock, Michael Doudoroff, Roger Stanier, Melvin Cohn, Aaron Novick, Seymour Benzer, Günther Stent et bien d'autres encore. Une franche gaité régnait autour de la table où fusaient force plaisanteries. Après le dîner, la soirée se terminait dans un café du Quartier Latin où l'on réformait le monde tout en vouant Lysenko aux gémonies.

A l'époque, je travaillais à l'Institut Pasteur de Garches et mon plus grand désir était de venir travailler dans le Service de Physiologie Microbienne où m'attiraient la qualité des hommes et l'intérêt des problémes étudiés. Malheureusement, l'exiguité des locaux du "grenier" ne permettait pas mon transfert et ce n'est qu'en 1954 qu'André Lwoff m'invita à rejoindre le groupe de Jacques Monod avec qui il partageait son Service. Durant les années qui précédèrent ma venue au laboratoire, André Lwoff avait été mon "parrain" au CNRS et je ne peux que me féliciter des conseils que j'ai reçus de lui au cours des entretiens qu'il m'accordait. C'est à son instigation que je me rendis en 1948 à Oxford, dans le laboratoire de D. D. Woods, pour m'initier aux principes de la nutrition microbienne; les quelques mois que j'y passais me furent éminemment profitables.

Quand j'arrivai dans le Service de Physiologie Microbienne, Jacques Monod et Melvin Cohn venaient de découvrir le caractère "gratuit" de l'induction de la β-galactosidase par les thiogalactosides. Ceci permettait une étude du "métabolisme de l'inducteur", expression fort prisée à l'époque. Jacques Monod me proposa de synthétiser le thiométhylgalactoside radioactif et de rechercher si l'on pouvait, à l'aide de cette molécule marquée, trouver à quelle macromolécule l'inducteur se liait au cours de l'induction. La question posée était hardie, sa solution théorique fut trouvée en 1958 par Jacob et Monod et confirmée expérimentalement en 1967 par Gilbert et Müller-Hill. En 1954, les concepts n'étaient pas assez mûrs ni les techniques assez raffinées pour que la solution fût trouvée. Mais ou cours de ces études, je fis cependant une découverte. Le thiométhylgalactoside radioactif était fortement concentré à l'intérieur des cellules de *Escherichia coli* si celles-ci avaient été préalablement préinduites, alors que chez les bactéries non préinduites, l'accumulation était négligeable. En dix jours, les caractéristiques essentielles du nouveau système étaient découvertes: compétition par d'autres thiogalactosides; stéréospécificité du transport; transformation du thiométhylgalactoside en une substance séparable par chromatographie (caractérisée par la suite comme étant le 6-acétylthiométhylgalactoside); identification des mutants cryptiques qui manquaient du nouveau système qui nous appelâmes galactoside-perméase; enfin, pléiotropie du caractére qui gouverne la constitutivité ou l'inductibilité des synthèses de la galactosidase et de la perméase.

Simultanément, j'étendis l'étude des perméases bactériennes aux amino acides et interprétai le mécanisme de certains antagonismes entre amino acides pour la croissance bactérienne qui restaient inexpliqués.

Mon séjour chez André Lwoff fut de courte durée; mais en allant travailler en 1955 dans le nouveau service de Biochimie Cellulaire dirigé par Jacques Monod, je n'étais séparé de la Physiologie Microbienne que par deux étages. Les membres des deux services se retrouvaient à l'heure du déjeuner dans un réfectoire improvisé dans un couloir du service.

Ces quelques années me marquèrent profondément et je devais garder pour l'Institut Pasteur en général et pour André Lwoff en particulier un attachement et une gratitude facilement explicables.

En 1959, les huit mètres carrés que j'occupais chez Jacques Monod étaient devenus trop exigus pour la réalisation d'un programme relatif à la régulation d'une voie de biosynthèse que j'avais commencé à étudier à Garches en 1951 et que je poursuis encore.

Les études d'André et Marguerite Lwoff sur la nutrition des microorganismes montraient une variation extrême dans les besoins en facteurs de croissance de diverses espèces; avec les travaux de Garrod sur les "erreurs innées du métabolisme", de Haldane sur la synthèse des pigments des fleurs, de Knight et Fildes sur les bactéries, de Beadle et Ephrussi sur les pigments des yeux des drosophiles, elles préfigurent l'essor de la génétique biochimique dont le succès est dû à l'obtention de mutants dont les lésions reproduisaient les phénotypes rencontrés dans la nature. Beadle, puis Davis utilisèrent *Neurospora crassa* et *Escherichia coli* pour élucider les principales voies de biosynthèse des métabolites essentiels. J'étais très attiré par ces questions et mon incursion dans le domaine de la perméabilité cellulaire ne devait être qu'un épisode. Après avoir contribué à Garches à l'élucidation de la voie de biosynthèse de la thréonine chez *E. coli,* je décidai d'en étudier la régulation.

En 1968, André Lwoff prit une retraite anticipée; "retraite" purement théorique, car s'il quitta l'Institut Pasteur, ce fut pour diriger l'Institut du Cancer du Centre National de la Recherche Scientifique à Villejuif. Mes amis Jacques Monod, Élie Wollman et François Jacob me pressèrent de poser ma candidature à sa succession à la tête du Service de Physiologie Microbienne. Je n'hésitai pas un seul instant.

J'écris ces lignes sur le bureau d'André Lwoff, assis dans son fauteuil. Je mesure le risque d'une telle succession. Ce laboratoire et

ses rejetons ont été une pépinière de découvertes qui ont jeté les fonda-
tions de la biologie moderne, pour n'en citer que quelques-unes: rôle
coenzymatique des facteurs de croissance, lysogénie, circularité du
chromosome bactérien, répresseur cytoplasmique, opérateurs et opérons,
ARN messager, théorie des transconformations allostériques. L'héritage
est lourd; il faut essayer de l'assumer.

La belle époque

FRANÇOIS JACOB

"L'art du chercheur, me dit un jour André Lwoff, c'est d'abord de se trouver un bon patron". Il m'est difficile de ne pas m'associer à une telle déclaration. La guerre finie, après avoir tenté les métiers les plus divers, je décidai de m'essayer à la biologie. Par suite des circonstances, j'avais bâclé mes études de médecine. J'ignorais tout de la médecine comme de la biologie. Je savais pourtant, par quelques amis, qu'il existait à l'Institut Pasteur un laboratoire de qualité exceptionnelle, celui que dirigeait André Lwoff et où travaillait Jacques Monod. Connaissance à la fois précieuse et rare, car si ce laboratoire était déjà fameux dans le monde entier, il restait bien souvent ignoré en France, en particulier des étudiants à qui personne n'enseignait la microbiologie. Je décidai donc d'apprendre la biologie en travaillant là. Quelques années auparavant, quand je me livrais aux joies de l'industrie pharmaceutique, j'avais eu l'occasion de rencontrer une fois André Lwoff et une fois Jacques Monod. Plus jeune, ce dernier m'intimidait moins. Mais à ma demande de travailler avec lui, Monod répondit: "Je n'ai pas de place. Et d'ailleurs, ce n'est pas moi le patron ici. Allez voir Lwoff". Ce que je fis.

Quelque peu ému, j'arrive au rendez-vous fixé à deux heures, un jour de septembre 1949. Je monte au grenier et j'emprunte pour la première fois le long couloir où les appareils de toutes sortes vous obligent à une série de méandres. André Lwoff achève son déjeuner dans son laboratoire en compagnie de deux charmantes personnes: Gisèle, sa secrétaire, et Antoinette, son aide-technique. Dans son bureau, je lui expose mes désirs, mon ignorance, ma bonne volonté. Il me fixe longtemps de ses gros yeux bleus, hochant la tête et me dit: "Ce n'est pas possible, je n'ai pas la moindre place".

A cette époque, je conservais encore certaines des vertus que m'avait inculquées la vie militaire. Notamment une persévérance qui confinait à l'entêtement. Nullement découragé par ce refus, je reviens au cours de l'hiver voir André Lwoff chaque mois. Et chaque mois, c'est le même regard bleu et doux, le même hochement de tête, le même refus. Vers le printemps, mon optimisme commence à s'entamer. Faute de grives, il me faut songer aux merles. Je me mets en quête d'un autre laboratoire que je trouve, mais préfère ne pas avouer. Pour être en règle avec moi-même, je risque au mois de juin une dernière tentative auprès d'André Lwoff, convaincu qu'elle aussi restera sans succès. En arrivant, je trouve l'oeil plus bleu, plus brillant que d'habitude, le hochement de tête plus profond, l'accueil plus chaleureux. Sans même me laisser le temps de lui exprimer à nouveau mes désirs, mon ignorance, ma bonne volonté, il annonce: "Vous savez, nous avons trouvé l'induction du prophage!". Je fais un "Oh! oh!" dans lequel je mets autant d'admiration que je le puis, tout en pensant "que diable est-ce donc qu'un prophage? Que peut bien vouloir signifier: induire un prophage? De quoi parle-t-il?". Il ajoute alors: "Cela vous intéresserait-il de travailler sur le phage?". Je bredouille un: "C'est juste ce que je souhaite.—Eh bien, venez le premier septembre". Je redescends l'escalier à la fois ravi et désorienté, un peu comme l'enfant qui obtient enfin le jouet longtemps convoité. Et surtout, je commence à ressentir mes premières inquiétudes à l'égard du phage. Il me faut consulter, dans la première librairie venue, un dictionnaire qui, bien entendu, ne me dira rien. Aujourd'hui encore, j'ignore pourquoi André Lwoff m'a admis dans son laboratoire. Peut-être à cause de mon entêtement. Ou de ma formation rien moins que classique. Ou encore de son enthousiasme provoqué par l'induction du prophage. Ce que je sais, c'est qu'à sa place, je n'aurais sûrement pas accepté dans mon service un individu de mon genre.

Au jour dit, j'arrive dans le service pour y faire mes premiers pas. C'est au mois de septembre 1950. Il règne une chaleur indécente dans le grenier. En l'absence de Monod, en voyage aux États-Unis, on m'installe dans son bureau, petite pièce munie d'une paillasse. En y entrant, j'ai la surprise d'y trouver, en train de sécher sur une ficelle, une culotte de femme, deux soutiens-gorge et une paire de bas. Le lendemain j'éprouve moins d'étonnement quand, du lavabo entièrement bouché, Mazé, le technicien du laboratoire, extrait péniblement des fragments d'une combinaison. Ce n'est, me dit-on, que la lessive de Catherine Fowler, jeune étudiante américaine venue faire un stage au laboratoire. Cela ne doit pas, ajoute-t-on, m'empêcher de faire de la physiologie microbienne. André Lwoff me donne pour mission, et pour sujet de thèse, la lysogénie chez *Pseudomonas pyocyanea*. Pendant l'été, j'ai eu le temps de donner un contenu au mot "phage". De mes lectures, je construis un petit univers, peuplé de bactéries, de phages et d'enzymes, sur lequel règnent quelques héros et, au premier rang, Max Delbrück et Salvatore Luria. Je m'imagine le premier comme trapu et bedonnant, le second long et sec. Sancho Pança et don Quichotte. Ce sera l'une de mes plus grandes surprises de découvrir que la réalité ne correspond pas à la représentation que je me suis faite.

Par cette chaleur accablante, je laisse ouverte la fenêtre pour repiquer mes souches de *Pseudomonas*. Brusquement s'entrouvre la porte et apparaît une tête. Une voix à l'accent chantant dit: "Ah! vous ensemencez les bactéries avec la fenêtre ouverte, vous! Il ne faudra pas vous étonner si vous faites beaucoup de découvertes avec des contaminants". La porte se referme. Et c'est ainsi que je fais la connaissance d'Annamaria Torriani. Mais si mon entrée dans ce petit univers ne se fait pas sans surprise, si je me trouve tout d'abord quelque peu décontenancé, ce n'est pas pour longtemps. Parmi le groupe qu'ils dirigeaient, André Lwoff et Jacques Monod avaient su faire naître une atmosphère où se mêlaient l'enthousiasme, la lucidité intellectuelle, le non conformisme et l'amitié. Le grenier était bien étroit, tout encombré de frigidaires, d'étuves et de centrifugeuses. Mais l'ardeur y régnait. A peine réalisée, à peine même projetée, chaque expérience était discutée sans fin, mise en pièces dans le couloir, refaite aussitôt. A tout instant, le débutant pouvait faire appel aux suggestions et aux critiques d'André Lwoff, à l'imagination de Jacques Monod, à la sagacité d'Élie Wollman, de Louis Siminovitch, de Melvin Cohn ou de Germaine Cohen. Grâce à la présence de la seule grande table du laboratoire, la pièce où je travaillais hébergeait les déjeuners. Là se poursuivaient sans fin les palabres sur les idées ou les techniques. Marguerite Lwoff et Sarah Rapkine y apportaient une note moins grave et veillaient à la cérémonie du café. Rarement la férocité intellectuelle s'est-elle mêlée aussi étroitement à l'indulgence et à la cordialité que pendant ces repas.

Toute cette effervescence n'allait pas sans quelques surprises pour le novice. Pendant un an, j'ai travaillé dans la même pièce que Seymour Benzer. Celui-ci était venu pour un an régler son compte à l'adaptation enzymatique en la traitant par le phage. Il avait un horaire très personnel. Arrivant pour déjeuner de pis de vache ou de testicules de taureau, il travaillait surtout la nuit et insistait pour avoir alors accès à la bibliothèque de l'Institut Pasteur; ce qui s'accordait mal avec les moeurs et la tradition de ce lieu vénérable. Le matin en arrivant, il laissait tomber un "Hie". A 16 heures 17, il émettait la plus sonore des éructations, et c'était tout. Après cette longue période d'incubation, nous sommes devenus d'excellents amis. Je me souviendrai aussi pendant longtemps du premier séminaire auquel il me fut donné d'assister. Toute la fleur de ce qu'on appelait alors "l'adaptation enzymatique" était là. J'eus

A close one.

BIO-AGRICULTURAL LIBRARY
UNIVERSITY OF CALIFORNIA
RIVERSIDE, CALIFORNIA 92502

l'impression d'assister à une corrida, dans le bureau d'André Lwoff, promu pour la circonstance au rang d'arène. L'orateur, Sol Spiegelman, jouait le rôle de taureau; Jacques Monod celui de toréador; Roger Stanier et Martin Pollock, de piqueurs de banderilles. Melvin Cohn à qui était dévolu le rôle de picador avait l'art de vriller sa lance au bon moment. Campé devant le tableau noir, Sol jouait avec un morceau de craie comme dans les films d'avant guerre George Raft avec une pièce de monnaie. Il décrivait les mérites des particules cytoplasmiques, chargées d'apparaître ou de disparaître pour rendre compte de l'adaptation enzymatique chez la levure. A peine une expérience décrite, le picador intervenait, les aides effectuaient quelques passes et le toréador réglait son compte à l'expérience. La tête enfoncée dans les épaules, Spiegelman se ramassait, relançait sa craie en l'air et décrivait une nouvelle expérience, aussitôt mise en pièces. Ceci pendant près de deux heures. Derrière son bureau, André Lwoff présidait avec le sourire et comptait les coups. A chaque passe, les spectateurs étaient sur le point de crier "Ole". La mise à mort eut lieu au café voisin.

J'ai eu la chance d'arriver à un moment exceptionnel dans le grenier de l'Institut Pasteur. Au moment où naissait une science nouvelle. Où la biologie en pleine effervescence, changeait ses modes de pensée, découvrait dans les microorganismes un matériel neuf et simple, se rapprochait de la physique et de la chimie. Un moment où il suffisait de travailler, de faire des expériences pour se retrouver en terrain inexploré. Deux pôles rayonnaient alors sur le couloir du grenier. A un bout, André Lwoff et son équipe déclenchaient le développement du phage en déversant sur les cultures de bactéries lysogènes du rayonnement ultraviolet, de l'ypérite, de l'eau oxygénée, et tout ce qu'ils pouvaient trouver en fait de substances mutagènes ou cancérigènes. A l'autre extrémité, Jacques Monod et son groupe déclenchaient la synthèse de la β-galactosidase en ajoutant aux cultures de bactéries ML, extraites de l'intestin d'André Lwoff, les β-galactosides les plus raffinés où étaient substitués tour à tour l'un ou l'autre des radicaux. Chacun "induisait" donc à sa façon, convaincu que les deux phénomènes n'avaient rien en commun, sinon un mot.

J'ai eu le privilège de contribuer à établir qu'André Lwoff et Jacques Monod se trouvaient tous deux dans une situation semblable à celle de Monsieur Jourdain: sans le savoir, ils faisaient la même chose, c'est-à-dire qu'ils passaient leur temps à inhiber l'inhibiteur. C'est, en effet, la juxtaposition des deux systèmes, le phage et la β-galactosidase qui a donné à l'analyse son efficacité et sa valeur. Avec Élie Wollman, nous avions fait de la conjugaison chez *E. coli,* un outil permettant d'étudier le matériel génétique de la cellule bactérienne et son fonctionnement. Il donnait le moyen d'analyser aussi bien la régulation du prophage que celle du lactose. Dans les deux systèmes, l'utilisation des mêmes méthodes conduisait aux mêmes résultats, aux mêmes concepts, au même langage. Toute propriété trouvée dans un cas se retrouvait dans l'autre. Toute mutation qui était obtenue ici d'abord, était isolée là ensuite. Toute conclusion tirée du phage conduisait à des prédictions pour la β-galactosidase et inversement. Rarement mariage a aussi étroitement uni le coeur et la raison.

Les années que j'ai passées dans ce grenier à apprendre mon métier demeureront parmi les plus fécondes et les mieux remplies de mon existence. Dans le travail de chaque jour, dans les discussions toujours recommencées avec André Lwoff, j'ai trouvé les critiques et les conseils nécessaires au débutant; j'ai trouvé aussi, chez lui, et chez Marguerite Lwoff, la chaleur humaine, peut-être plus nécessaire encore. L'étonnante jeunesse d'esprit d'André Lwoff le portait toujours à accueillir avec intérêt le résultat le plus étrange que pouvait lui apporter le novice, à s'enthousiasmer pour une idée un peu nouvelle. Il est assez exceptionnel en biologie de pouvoir, à nombreuses reprises, changer de domaine, de matériel, d'attitude. Il est plus exceptionnel encore de réussir à chaque fois et de se retrouver toujours là où il se passe quelque chose, quelle que soit l'orientation nouvelle. Tour à tour protozoologiste, microbiologiste, physiologiste, généticien, virologiste et cancérologue, André Lwoff a véritablement contribué au premier rang à la fondation de plusieurs sciences. Mais ni la renommée, ni les prix, ni les académies, ni même le manque d'académie, n'ont entamé son sens critique, son humour, sa bienveillance.

Malgré son rôle dans l'évolution de la biologie, André Lwoff a choisi d'avoir peu d'élèves. Ce qui l'a toujours intéressé, c'est de travailler lui-même, non de régner sur un ou plusieurs empires. J'ai eu la très grande fortune d'être l'un de ces rares élèves. A ma naissance à la science, il a guidé mes premiers pas avec une bienveillance et une amitié qui perçaient toujours derrière son ironie. Plus tard, quand ma démarche s'est assurée, quand j'ai commencé à m'agiter et à remuer plus d'air, il n'a jamais cessé de m'encourager, de m'aider, de me pousser en avant. Mais la plus haute marque de confiance qu'il m'ait jamais donnée fut de me nommer à l'une des plus importantes fonctions de laboratoire: celle de membre du Comité de Nomenclature qu'il présidait. J'ai eu ainsi l'honneur de participer à l'élaboration de quelques trente ou quarante mots entièrement nouveaux, tous par la combinaison de racines grecques. Trois ou quatre ont survécu. Dans ce domaine, dix pour cent représente un rendement non négligeable. C'est récemment que j'ai réalisé pourquoi André Lwoff attachait un tel prix à ces problèmes. Dans une discussion entre Confucius et un élève, celui-ci demande, bêtement, au Maître pourquoi, dans les actes de la vie, il insiste tellement sur l'importance des noms. Et Confucius de répondre:

"Si les noms ne sont pas corrects, le langage n'est pas en accord avec la vérité des choses. Si le langage n'est pas en accord avec la vérité des choses, les affaires ne peuvent être traitées avec succès. Si les affaires ne peuvent être traitées avec succès, les bonnes manières et la musique ne s'épanouiront pas. Quand les bonnes manières et la musique ne s'épanouissent pas, les punitions ne sont pas convenablement distribuées. Quand les punitions ne sont pas convenablement distribuées, les peuples ne savent remuer ni le pied, ni la main".

Research on lysogeny at the Service de Physiologie Microbienne, 1954–1956

DALE KAISER

In 1954 André Lwoff ceased his work on lysogeny to begin the study of poliovirus multiplication. Even though I had come to the Service de Physiologie Microbienne at just that time to do postdoctoral work on lysogeny with François Jacob, André Lwoff's influence on me was strong for several reasons. In the first place, the Service de Physiologie Microbienne was Lwoff's creation. He had brought together a remarkable staff which included Élie Wollman, François Jacob, Pierre Schaeffer, Jacques Monod, and George Cohen. Also, Lwoff in his review "Lysogeny," published in *Bacteriological Reviews* (1953), had set forth two ideas which have dominated research in the field since that time. He presented both ideas with compelling clarity and beauty of expression. One was the idea of *prophage,* the genetic material of a temperate bacteriophage reduced to an inactive state in which it replicated in synchrony with its host bacterium but did not direct the synthesis of new virus particles. The other was that prophage could be brought from this inactive state back to an active, virus-particle-producing state by various treatments called *induction*. The aim of this essay is to give a brief account of the work on lysogeny in Lwoff's Service during the 1954–1956 period.

In 1953 Élie Wollman had discovered, as had E. and J. Lederberg independently, that prophage λ was part of the bacterial chromosome located near a cluster of genes involved in the conversion of galactose to glucose. Under Lwoff's supervision François Jacob had just completed his thesis on lysogeny in *Pseudomonas pyocyanea,* in which he had particularly studied the phenomenon of induction. Then, working together, Jacob and Wollman observed that development of prophage λ is induced when a lysogenic male is crossed to a nonlysogenic female (Jacob and Wollman, 1954). This was the first demonstration that prophage is held in an inactive state by a system of negative control.

Thus in late 1954 and early 1955 Jacob and Wollman were trying to dissect the mechanism of bacterial recombination into its component steps. To discriminate between formation of mating pairs and subsequent gene transfer they separated the partners by strong shearing forces in a Waring Blendor. The experiment could have been done with a single genetic marker, because the prevailing idea at the time was that sex in bacteria, like sex in all other organisms, involved transfer of a complete nucleus. But they chose to use two markers: "classical" auxotrophy for threonine-leucine and their newly discovered zygotic induction as an internal control. To everyone's surprise, the two markers showed different kinetics of transfer (Wollman and Jacob, 1955).

Interesting new results, such as this one, were often discussed at morning tea, served in the narrow corridor outside Lwoff's office by his associate Sarah Rapkine. An explanation for the surprising transfer kinetics was arrived at over tea one day by Wollman, Jacob, and Monod. They called it the "spaghetti hypothesis" because it suggested that the chromosome of the male was transferred to the female like a long piece of spaghetti. Separating the two cells before completion of transfer breaks the chromosome. Then only genes near the leading end have any chance to be incorporated into the chromosome of the zygote.

One experiment confirming the spaghetti hypothesis involved the measurement of killing of recombinants by ^{32}P decay in zygotes. Clarence Fuerst quietly carried out these experiments several doors down the hall from the laboratories of Wollman and Jacob. At the same time Tom Anderson and René Mazé in another part of the laboratory had begun to isolate conjugating pairs of bacteria microdrops in order to follow segregation from single cells (Anderson, 1958). They used the same technique (and possibly the same micromanipulator) that Lwoff and Gutmann (1949) had used to demonstrate that the spontaneous production of bacteriophage by single *Bacillus megaterium* cells was always accompanied by their lysis.

Sometimes enthusiasm for science spilled over into practical jokes. A visit of Haile Selassie, *le roi d'Ethiopie,* to the Institut Pasteur was the occasion for one. About 11 o'clock on the morning of his visit Jacob breathlessly told Wollman that the Emperor wanted to see bacteriophage plaques and that he, Wollman, was to show him some at noon. Because it was a state visit, Jacob told Wollman, the Institut Director wanted him to be dressed in traditional Institut garb—white apron and tight, pillbox-shaped hat. It took Wollman an hour to locate his props, and he was too rushed to consider the plausibility of the arrangement. At 12 o'clock he was just ready, standing at the street properly dressed and holding two plates with plaques, but the Emperor only drove past the Institut in his car.

Lunch, which everyone in the laboratory ate together, was usually reserved for discussions about art (a special interest of Lwoff's), people, and politics. Élie Wollman read *Le Monde* from cover to cover and

The coffee ceremony.

remembered everything he read. He seemed to know the family and political biography of every French politician. Of the American post-doctoral fellows in the Service at that time only Cyrus Levinthal had the courage to enter into these arguments.

Though Lwoff had stopped work on bacteriophage in 1954 and had started to study animal viruses, he maintained a strong interest in induction, lysogenization, and immunity. During the previous year Lwoff (Lwoff, Kaplan, and Ritz, 1954) had explored the "multiplicity effect" in the lysogenization of *Salmonella typhimurium* by phage A, whereby the frequency of lysogenization increased rapidly with multiplicity in the range 1–10. That same year Lieb (1957), in Lwoff's laboratory, found that a clear mutant (a *cI*-type clear) could give the multiplicity effect as long as the bacteria also received at least one coinfecting c^+-phage particle. Because these experiments and the problems they raised were still fresh in his mind, I suspect, Lwoff was interested in experiments I was carrying out on clear mutants of phage λ. Three of the genes Jacob and Wollman (1954) and I had used to map the λ chromosome strongly affected the frequency of lysogenization. When a phage was mutant in one of these genes, it gave clear plaques. I isolated a number of independent mutants and mapped and complemented them with each other (Kaiser, 1957). Complementation experiments in which two different clear mutants were allowed to infect the bacteria at different times particularly interested Lwoff, and he translated into French a short note describing them.

By 1955 Jacob and Wollman had accumulated many different temperate bacteriophages from strains of *Escherichia coli* isolated in various Paris hospitals. The phages were selected for ability to grow on *E. coli* strain K12 and for differences in immunity specificity. Jacob suggested that I try crossing them with λ. One, called 434, proved to be the most revealing, because it recombined with λ everywhere except in the vicinity of the *cI* gene. The properties of a recombinant called 434 hybrid showed that the specificity of immunity is determined by genetic material in the vicinity of the *cI* gene (Kaiser and Jacob, 1957). This hybrid was generated by five successive backcrosses to λ, always selecting for 434 immunity. Subsequent work has shown that the *cI* gene itself and neighboring operator genes determine immunity specificity.

Leaving the Institut Pasteur in the fall of 1956, I was anxious to start my first job. I look back on 1954–1956, however, as one of the most intellectually exciting periods in my life, and I am grateful to André Lwoff and his colleagues for making it so.

REFERENCES

Anderson, T. F. (1958). Recombination and segregation in *Escherichia coli. Cold Spring Harbor Symp. Quant. Biol.,* 23, 47.

Jacob, F., and E. L. Wollman (1954). Étude génétique d'un bactériophage tempéré d'*Escherichia coli.* I. Le système génétique du bactériophage λ, *Ann. Inst. Pasteur,* 87, 653.

Jacob, F., and E. L. Wollman (1958). Sur les processus de conjugaison et de recombinaison chez *E. coli.* IV. Prophages inductibles et mesures des segments génétiques transférés au cours de la conjugaison, *Ann. Inst. Pasteur,* 95, 497.

Kaiser, A. D. (1957). Mutations in a temperate bacteriophage affecting its ability to lysogenize *Escherichia coli, Virology,* 3, 42.

Kaiser, A. D., and F. Jacob (1957). Recombination between related temperate bacteriophages and the genetic control of immunity and prophage localization, *Virology,* 4, 509.

Lieb, M. (1957). Participation of a non-temperate phage in lysogenization, *Ann. Inst. Pasteur,* 93, 52.

Lwoff, A. (1953). Lysogeny, *Bact. Rev.,* 17, 269.

Lwoff, A., and A. Gutmann (1949). Production discontinue de bactériophages par une souche lysogène de *Bacillus megatherium, Compt. rend.,* 229, 679.

Lwoff, A., A. S. Kaplan, and E. Ritz (1954). Recherches sur la lysogénisation de *Salmonella typhimurium, Ann. Inst. Pasteur,* 86, 127.

Wollman, E. L., and F. Jacob (1955). Sur le mécanisme du transfert de matériel génétique au cours de la recombinaison chez *Escherichia coli* K12, *Réud. Acad. Sci. Paris,* 240, 2449.

From lysogeny to animal viruses

RENATO DULBECCO

In the summer of 1962 I attended the Symposium at Cold Spring Harbor and reported on the work of my laboratory on the growth of animal viruses in tissue culture and the new developments stemming from it. Among those attending the symposium was André Lwoff. Toward the end of the meeting, he asked whether he could come to Pasadena and learn about tissue culture and animal viruses. I was very happy: it was a good omen for the future that my first convert was the recognized master of microbiology and the father of lysogeny.

Several months later André and Marguerite Lwoff arrived at the California Institute of Technology. There I had several rooms located at various places in the basement of the Kerkhoff Laboratory. George Beadle agreed to give me another room, at the end of the corridor, to accommodate the guests.

We had recently developed a system employing poliovirus and cultures of monkey kidney cells. At that time, the most basic aspects of the reproduction of the virus in the new system were being investigated. The work was stimulated, on one hand, by the great excitement coming from the phage work, to which I also had contributed previously and which was going on full steam at Caltech under Max Delbrück's guidance, and, on the other hand, by the pressing demand for better understanding of animal virus biology, and of poliovirus especially, since the National Foundation for Infantile Paralysis was expending its vast resources in a quest for the prevention of poliomyelitis.

At that time, there was a vast discrepancy between our meager understanding in the animal virus field and the tremendous progress of phage work, but I felt very enthusiastic because animal virology had started yielding to our efforts. The success of our phageologist friends only paved the way and set the goals for our own evolution. The important thing was to ask the right question.

However, there was still a big gap between what we *wanted* to do and what we *could* do. We were acutely aware that progress was strictly dependent on the available means of investigation, and that the problems chosen must take full advantage of our technology but would be limited by it also.

So when André and Marguerite Lwoff arrived in Pasadena, they, Marguerite Vogt, and I sat down to decide what to do. (We were first faced with a practical problem of two Marguerites in the group; each one tended to mark her materials with an M, and this led to confusion. The problem was readily solved when Marguerite Vogt announced that her middle name also began with M, and that therefore she would henceforth be known as MM, leaving the single M to Marguerite Lwoff. This worked so well that even to this time there is glassware in our laboratory indelibly marked MM.)

We had been preoccupied with the problem of virus release from infected animal cells. We knew that phages are released in sudden bursts, whereas encephalomyelitis virus, which I had previously studied in collaboration with Marguerite Vogt, is released slowly, one, or a few, virus particles at a time. We had found this out by a laborious experiment in which the technique of the statistical "single-burst experiment," classical in phage work, had been transplanted to the animal cell system, taking advantage of some characteristic and favorable properties of these cells. We were interested in studying the release of poliovirus. André suggested that we try the technique he had developed in the study of lysogeny, that is, to pick individual cells under direct visual control of the microscope, and transfer them into small drops of medium under paraffin oil. This would avoid the uncertainties and greater complexities inherent in the statistical approach. The experiment would be easier with animal cells because they are much larger than bacteria. The idea was excellent: it offered the opportunity to introduce a new element of technology into the developing field of animal virology, and everyone had the opportunity to learn something new.

The project involved some engineering problems. We could borrow a De Fombrunne micromanipulator from Max Delbrück (he had seen the potential of the approach in bacteriophage work and had bought the instrument). However, we had no chambers and no microforge. Undaunted, we proceeded to build them. The success of these enterprises has been recounted by André Lwoff himself, and I am not going to repeat the story.

Isolating single animal cells from a suspension proved rather easy, since they could be seen at low magnification: they could be picked by hand. André was the maestro: he showed us how to make very fine pipettes, using an alcohol flame and watchmaker's forceps. He also told us that we must have a steady hand. The requirements, he said, were not to have smoked for twenty years, not to have had recent contact with any form of alcohol, and to have had a good night's sleep. As I later found out, I could work more successfully the morning after a big party, but I never told him.

Once the technical problems were over, the experiments proceeded fairly smoothly. It was fascinating to *look* at the cells and see their dramatic changes at the time virus was being released. All of us, being so used to working with essentially invisible phage and bacteria, were fascinated. I went as far as to take pictures of the cells as they were performing, a weakness which the maestro tolerated with benevolent amusement. By removing the medium of a drop at frequent intervals and then assaying it for virus, we could catch the process of release as it occurred in a single cell. We found that the whole process occurred quite rapidly, within 10 or 20 minutes, and was therefore very different from the release of encephalomyelitis virus. On the other hand, it was not quite a burst, as with phage. These differences gave us a good glimpse of the great diversity of animal virus biology. We were also amazed at the enormous asynchrony of the virus release in different cells. Of course, Cairns had already shown this for influenza virus, but at first we could not quite believe his results because they had been obtained in chicken eggs. Altogether, the experiment was very exciting, as it combined many interesting new methodological approaches and allowed insight into several aspects of animal virus reproduction.

The significance of the experiment extended considerably beyond its immediate results, as we have since utilized it many times for a number of problems. We simplified it by eliminating the use of the micromanipulator and replacing the special chambers with plain glass or plastic Petri dishes containing small drops of medium under paraffin oil. This technique allowed us, for instance, to fully characterize the "carrier" state of polyoma virus in mouse embryo cultures, and to distinguish it from a "lysogenic" state. More recently, it proved very useful in studying a line of 3T3 cells transformed by a temperature-sensitive mutant of polyoma virus, which constantly releases virus particles but is not in a carrier state. In this way, the study of virus release from individual cells has brought us over the years to cover the broad range of unenveloped and enveloped virulent viruses and that of moderate tumor-producing viruses.

In due time André and Marguerite Lwoff returned to their laboratory at the Institut Pasteur and immersed themselves in the study of poliovirus reproduction, which led them to a number of important accomplishments, such as the characterization of the thermosensitive step and the identification of the *rct* mutants, so important for Sabin's investigations of live polio vaccines.

Recently André and Marguerite returned to my laboratory. At this time, I was concentrating on studies of the mechanism of cell transformation by polyoma virus and SV40, and I wished to enlist André's experience and intuition in the study of this process, which has some similarity to lysogeny. Needless to say, André gallantly accepted the challenge posed by a system that was entirely new to him. Now endowed with the highest recognitions of the scientific world, he again took his place at the bench, close to our graduate students and postdoctoral fellows. It was, for these young people at the beginning of their careers, a most inspiring experience. I think it must also have been gratifying for André to recall old times and to see again some of the MM-marked glassware.

Again we drew on André's past experience with lysogeny, and we asked questions on the possible existence of a viral repressor in the transformed cells. The work initiated at that time is now being developed, both in my laboratory and in André's laboratory in Villejuif. In this work, the concepts derived from lysogeny, to which André Lwoff contributed so much, are helping to make possible a deeper exploration of control mechanisms in animal cells, fulfilling the anticipations promised by the elucidation of the control mechanisms operating in lysogenic cells.

La vie est dur

NEAL GROMAN

At the Second International Colloquium on Bacteriophage at the Abbé de Royaumont late in July 1958, it was obvious that Lwoff had a new hobby: he had shifted to animal virology and had discovered the inhibitory effect of high temperatures on poliovirus replication. He was thoroughly enjoying the mystery of his finding in his discussions with the participants. One couldn't help feeling that something significant lay behind the observation. After the beautiful work on lysogeny one had to conclude that Lwoff was endowed with magical powers which permitted him a dialogue with Nature denied to others.

Lwoff brought the same enthusiasm and sparkle to this new problem that he had to others. As a matter of philosophy he felt that Nature was not obvious, but that each of her manifestations was a reflection of something basic. Perhaps he, too, felt that he had captured a magical power, and that once again the path would lead to a major discovery. That it did not is Nature's loss, for she hardly has a more ardent and insistent proclaimer of her beauty than Lwoff.

We settled down to work after the August holidays. The physical setting of the Service de Physiologie Microbienne was historically fascinating, but unbelievably ancient to the eyes of a young American investigator. Lwoff had a single modest laboratory in which he worked and an adjoining office.

Our first basic agreement was on language. During the first few days there were some abortive efforts to discuss the work in French, but it quickly became apparent that this would not do. My French training was for reading and was unfortified by practice in spoken French, and in any event Lwoff's English was superior to my own. The outcome was obvious from the first, and English was spoken thereafter. I felt that this arrangement disappointed Lwoff. Fortunately our children went to French public school and made amends for their parents. I suspected that André saw their facility with the language as the greatest benefit derived from that year. That this was more than a suspicion was made obvious when we met in Moscow at the IXth International Congress of Microbiology in 1966. As we spoke briefly in the midst of a gigantic reception, one of the first questions he asked was, "How is the children's French?"

Sharing a laboratory can be a difficult business. It involves all the problems of sharing a household. The limiting factor in Lwoff's laboratory was a single hood in which the art of virology and tissue culture was practiced. The hood itself was a piece of mechanical wizardry, one of the many touches which showed the combination of a perceived need (Lwoff) with the artistry of a skilled mechanic (Mazé). The problem of sharing the hood was settled with the same amiability and consideration which was to mark our relationship throughout the entire year. In the end we had a shared time plan with easily worked-out hours and a common-sense regard for each other.

A problem did develop at one point. In my zeal to shut things down properly, I had the unconscious habit of tightening up all the taps quite hard. One day after an obviously difficult bout with a water faucet, Lwoff felt constrained to comment on this habit and asked what it was all about. I offered him the explanation that I had milked cows for many years as a boy and that this had left me with very strong hands. He appreciated the originality of the hypothesis, and accepted it with that marvelous sense of humor so characteristic of him.

Everyone who has been at the Pasteur will remember the lunches downstairs. This genial meeting of the Lwoff-Monod groups was a daily ritual. Its very occurrence was a reflection of Lwoff's capacity to encourage the growth and development of others. Although most of the conversation was in French, enough English was spoken that everyone could get some feeling of what went on. The disillusionment of the French scientists with the politics of their own land was readily apparent. (General de Gaulle had just ascended to the presidency and was in the process of extricating France from Algeria.) On the other hand they were not less critical of U.S. policies and actions, and those of us from the United States had our nationalism stirred to levels not reached since the Armistice Day parades of our youth.

A second ritual, afternoon tea, was observed later in the Lwoff laboratory. At that time tea was made with strict adherence to the English method of brewing, a striking concession to foreign technical expertise.

My discussions with Lwoff rarely penetrated to the personal level. Occasionally he would reminisce about the Institut scientists he had known who had added to its luster. His sense of history was very much alive, whether dealing with microbiology or more secular matters. Pasteur was literally across the street, and Leeuwenhoek in a nearby country. His description of the round-up of Jewish scientists at the Institut during the Nazi occupation cannot be forgotten. He often spoke of Jacob's stubborn and finally successful efforts to work at the Institut with him. It seemed to constitute something of a miracle in his eyes, one easily understood in the light of subsequent history.

I may be giving the impression that science was not in the air at the laboratory. The polio work was going on, and I was assigned to the problem of isolating "cold" mutants of poliovirus. In addition to the polio work, Lwoff continued to shape the virus classification scheme which set the pattern for future years. It was also a year in which Jacob and Monod were making great theoretical strides in their work, and Lwoff was actively engaged in the discussions. Jacob's laboratory was across the hall, and often the first hour of the morning saw the scientists conferring in Lwoff's office. On occasion Lwoff would emerge from such a conference visibly excited by the direction the work was taking.

In his daily work in the laboratory, Lwoff ran his own experiments. He set up the experimental flasks, took his own samples, and entrusted the virus assays to his wife, Marguerite. His mood in the laboratory was very stable. Occasionally I would hear the statement, "La vie est dur" escape his lips, but it seemed more a philosophical comment than one related to an immediate occurrence.

By year's end the "cold" mutants had been isolated, and the results were incorporated into an article written in French by the Lwoffs (Groman, A. Lwoff, and M. Lwoff, 1960). Although the scientific results of the year were modest, my sojourn at the Institut had great impact on my general view of science and subsequently affected the course of my research and teaching. Thus both my students and I owe much to the year I was privileged to spend with Marguerite and André Lwoff.

REFERENCE

Groman, N., A. Lwoff, and M. Lwoff (1960). Recherches sur un variant dit "froid" du virus de la poliomyelite, *Ann. Inst. Pasteur,* **98,** 351–359.

Un Brésilien à Pasteur

LUIZ H. PEREIRA DA SILVA

Le qualificatif d'étranger ne s'applique à aucun des membres de cette famille (les chercheurs). Nombre des ressortissants des nations dites étrangères sont d'ailleurs tellement français qu'ils nous ont révélé des aspects insoupçonnés de notre grammaire et de notre syntaxe.   A. LWOFF,

Colloque "Sur le métabolisme des microorganismes,"
Marseille, 1963

Je n'oublierai jamais l'embouteillage du 3 juin 1968: je repartais le lendemain pour le Brésil, après six ans passés à l'Institut Pasteur. Je voulais saluer André Lwoff personnellement avant mon départ. Il m'avait dit de passer chez lui après le dîner. Cependant à huit heures du soir, j'étais immobilisé, au milieu de la place du Carrousel, au Louvre. Cela me faisait vraiment de la peine de ne pas revoir André Lwoff. J'accordais une valeur symbolique à ce geste. Je voulais serrer la main qui m'avait amené à Pasteur.

Mais après un moment, j'ai compris que je n'avais plus le temps et qu'il fallait renoncer; j'ai abandonné la voiture sur un trottoir et j'ai cherché un téléphone.

—Monsieur Lwoff, je suis pris dans un terrible embouteillage et je crains de ne pas pouvoir passer chez vous. Je regrette beaucoup d'être obligé de vous saluer par téléphone.

—Cela ne fait rien. Je vous souhaite un bon voyage et un heureux retour.

—Merci, Monsieur, je vais rejoindre mes trypanosomes. . . .

—Bien entendu, on revient toujours à ses premiers amours. . . .
a-t-il ajouté en riant.

C'était le rappel d'un entretien que nous avons eu un jour dans son bureau. Je lui expliquais avec ardeur mon intérêt pour les trypanosomes et les possibilités que j'entrevoyais dans l'emploi de ce matériel pour les études de différentiation cellulaire. "Les biologistes, m'a-t-il dit, essayent toujours de rationaliser l'intérêt qu'ils portent à tel ou tel organisme. Au fond, la vraie raison qui les pousse, c'est l'amour qu'ils ont pour une bête ou l'autre. Je comprends fort bien votre intérêt pour les trypanosomes, car moi-même j'en ai été un jour amoureux. Maintenant, les années ont passé et mon amour est devenu plus platonique. Je m'intéresse aux virus".

Je suis rentré au Brésil le lendemain, mais je n'y suis pas resté. Et pour moi aussi, d'autres intérêts ont supplanté celui pour les trypanosomes, malgré ma décision première de leur revenir.

Cette décision, elle datait de 1960. A ce moment, je me battais pour obtenir une place au laboratoire de Physiologie microbienne de l'Institut Pasteur. Je travaillais sur le cycle évolutif des trypanosomes et j'avais décidé d'apprendre la génétique bactérienne avec Jacob et Wollman, chez Lwoff. Lwoff avait repoussé poliment une première tentative d'approche. Dans sa réponse à ma lettre de demande il me conseillait d'autres laboratoires en Europe et en Amérique mais moi je m'étais mis dans la tête d'aller à Pasteur. Pourquoi? Il y avait naturellement Paris. Ensuite, il y avait les expériences Pardee-Jacob-Monod qui m'avaient séduit. Mais la raison principale était la présence de Lwoff. Il avait été protozoologiste. Il pourrait me guider dans cette reconversion de protozoologiste à biologiste moléculaire que je me figurais difficile.

On a beaucoup parlé du grenier surpeuplé du troisième étage de l'Institut Pasteur, où travaillaient Lwoff, Monod, Wollman, Jacob, Schaeffer et tous leurs collaborateurs. Pour moi, voici comment l'exiguïté d'espace a joué en ma faveur.

Par manque d'installation, tous ceux qui travaillaient chez Lwoff devaient traverser le laboratoire d'à côté, celui de Grabar, pour atteindre la pièce que tout chercheur en bonne santé, et suivant une journée de travail honnête, doit fréquenter au moins deux fois par jour. Cela permettait aux chercheurs de chez Grabar de se croiser souvent avec les gens de chez Lwoff. Parmi les premiers, il y avait à l'époque un compatriote et ami, Nussenzweig, à qui j'avais demandé de guetter Lwoff et de lui glisser de temps en temps quelques paroles à mon sujet. Il a si bien et si souvent fait la chose qu'il a fini, je pense, par déclencher un réflexe chez Lwoff. Celui-ci a commencé à s'intéresser à moi et un jour, finalement, à la suite de je ne sais pas quelles combinaisons avec Jacob, j'ai reçu une lettre où on m'annonçait que j'étais accepté pour septembre 1962.

Quand je suis arrivé au labo de Physiologie microbienne, celui-ci avait déjà bourgeonné et Jacob avait son service de Génétique cellulaire. J'ai commencé à travailler avec Jacob sur le phage lambda. Lwoff avait quitté les bactériophages depuis longtemps. D'après lui, il y avait déjà beaucoup de monde sur ce sujet.

Et pourtant à cette époque, quelques dizaines de personnes seulement s'intéressaient à lambda, alors qu'actuellement plusieurs centaines de chercheurs s'acharnent sur le pauvre phage. Il serait intéressant, un jour, d'évaluer l'investissement en efforts personnels et en argent qui a été fait sur lambda. On arrivera sûrement à des chiffres de nombreuses années-chercheurs et de milliers de dollars par nucléotide. Et l'on pourra alors se demander, suivant les principes modernes de gestion, si l'investissement a été rentable.

Laissons cette tâche redoutable aux historiens futurs de la science. On pourrait cependant déjà essayer d'esquisser les secteurs de la biologie où le phage lambda aura été à l'origine d'importantes conquêtes. Dès à présent, on peut dire qu'elles ont été nombreuses: c'est le phage lambda qui a inspiré l'idée de répresseur, le modèle de régulation négative, l'idée d'épisome, le modèle de réplicon, les bases moléculaires de la recombinaison génétique, le rôle régulateur de certaines structures de la molécule de DNA. Mais il est curieux de constater que la contribution de lambda au déchiffrage de la structure du matériel génétique, du code, et des mécanismes de transcription et de traduction est négligeable à côté des contributions de T_4 ou d'*Escherichia coli*.

L'historien pourra en déduire que lambda a servi principalement dans les secteurs abordés par des moyens plutôt biologiques que physiques ou chimiques et où la manipulation des organismes par des techniques microbiologiques et génétiques ont eu le rôle dominant. L'historien pourra ensuite se demander si certains scientifiques ont eu une influence personnelle, et dans le choix d'un tel organisme, et sur les voies de développement des travaux. Je crois qu'il lui sera facile de répondre par l'affirmative en constatant combien un chercheur comme Lwoff a joué un rôle déterminant par ses études sur la lysogénie.

A l'origine des énormes progrès réalisés par la Biologie après 1945, on peut distinguer certains éléments essentiels: l'introduction de nouvelles techniques physiques et chimiques; la participation d'un nombre important de mathématiciens et physiciens qui ont apporté un nouveau mode d'analyse des phénomènes biologiques, enfin la convergence de certaines disciplines biologiques, telles la cytologie, la génétique, la microbiologie et la biochimie.

C'est sur ce dernier point que certains rares biologistes comme André Lwoff ont apporté une contribution décisive en définissant les problèmes biologiques généraux à exploiter. Ils ont ainsi établi les liens historiques entre l'étape de synthèse en Biologie représentée par le Naturalisme du XIXe siècle, la Biologie analytique de la première partie du XXe siècle et la nouvelle synthèse représentée par la Biologie moléculaire.

Il est évident que pour déchiffrer les problèmes biologiques que les hommes de sciences abordent actuellement, telles que la différentiation cellulaire, la régulation du métabolisme dans les organismes supérieurs et l'activité psychique des animaux et de l'homme, on devra faire appel à de nouvelles techniques et à la participation d'un plus grand nombre de spécialistes dans différents domaines scientifiques.

Le travail de prospection sur ces nouveaux terrains se fait avec une intensité croissante. Cependant, il est aussi évident qu'il reste à énoncer les idées maîtresses et les nouveaux modèles expérimentaux à employer.

Pour ce faire, il faut espérer comme à l'étape précédente, en des biologistes possédant à la fois vaste culture et esprit indépendant. La Biologie moléculaire après son "âge d'or" est peut-être tombée déjà dans les mêmes vices de spécialisation qui ont perdu la Biologie d'avant-guerre et ses taxonomistes.

Pour sortir de cette impasse, on en est encore à attendre les vrais savants qui joueront un rôle équivalent à celui de Lwoff dans la naissance de la Biologie moléculaire.

Quarante degrés à l'ombre

MARC GIRARD

C'est à la suite d'un heureux hasard que je rencontrai pour la première fois Monsieur Lwoff. J'avais fait acte de candidature pour une bourse de formation à la recherche scientifique, la virologie m'attirait, je sollicitai donc une entrevue qu'il voulut bien m'accorder à Pasteur. De ce premier contact, je garde le souvenir d'un regard bleu et vif, d'un sourire prompt, mais d'un homme très intimidant. Il laissait volontiers, et peut-être à dessein, le silence s'installer dans la conversation, et me dévisageait alors, les mains jointes à hauteur du visage, d'un coup d'oeil à la fois chaleureux et narquois. Il voulut bien m'accepter pour filleul, et son premier soin fut de m'envoyer apprendre mon métier chez James Darnell aux U.S.A.

Deux ans plus tard, je le revis à New York lorsqu'il y vint invité par l'Albert Einstein College of Medicine. Il vint partager le déjeuner dominical de ma famille, et mangea, à cette occasion, son troisième plat de gigot en deux jours! Nous habitions alors New Rochelle, faubourg Nord-Est de New York, dont le nom avait une saveur assez pittoresque pour qu'André ait apporté avec lui crayons et calepin, dans l'espoir de croquer quelque scène marine. Las, de New Rochelle, seul le nom était pittoresque, et je me sentis presque confus de n'avoir à offrir au peintre que des bords de mer encombrés de détritus.

C'était pour m'initier à la biochimie du poliovirus qu'il m'avait envoyé aux États-Unis, or j'avais travaillé jusque-là sur la synthèse des ribosomes dans la cellule HeLa. Il me fit sentir la férule, et me convainquit de revenir dans le droit chemin de la virologie. C'est grâce à cette intervention que je débutai, sous la direction de Jim Darnell, les premières expériences de marquages très brefs à l'uridine qui devaient nous permettre de déterminer le temps nécessaire à la synthèse d'une molécule de RNA viral, et de mettre en évidence les étapes intermédiaires de la réplication du poliovirus. Je partais peu après, avec la bénédiction d'André, poursuivre ces recherches au Salk Institute à La Jolla.

De retour du Salk en juillet 66, je vins lui rendre ma première visite et voir le labo. Je retrouvai non sans panique le célèbre couloir du 3ème étage qu'il fallait emprunter, en haut de l'escalier d'honneur, pour se rendre dans le service de Physiologie Microbienne. Zigzaguant dans le service du Dr Grabar, ce couloir tenait du labyrinthe. Il était de plus encombré des matériels les plus hétéroclites, obstrué à maints endroits par centrifugeuses, étuves ou congélateurs, bordé ailleurs de deux rangées de placards en bois à la peinture défraîchie, qui servaient à leur sommet d'étagères où, faute de place, on stockait cartons, caisses ou bains-marie hors d'usage. Le bureau d'André occupait le fond de ce boyau, celui de François Jacob un appendice sur la droite, celui d'Élie Wollman, à gauche, un diverticule vitré où l'on pouvait à peine se tenir debout à cause de la pente du toit. Le labo d'André se résumait en fait à trois pièces, outre les laveries, étuves et chambre froide. Dans la première, à côté de son bureau, se trouvaient Marguerite Lwoff et Jacqueline Pagès; c'est là qu'André lui-même, jusqu'à ce jour, avait coutume de manipuler. La deuxième pièce, très exiguë, toute en longueur, était située à l'autre extrémité du couloir; elle abritait une boîte à gants, un microscope et divers bains-marie. Dans la troisième pièce, on trouvait le bureau de Danièle Ménard, qui devait bientôt devenir Madame Bucchini. Mais comme cette pièce servait aux titrages du poliovirus, les jours de titrage on en chassait Danièle, et deux techniciennes s'y enfermaient à clef sous prétexte d'éviter les contaminations microbiennes ou fongiques qu'auraient pu amener les courants d'air. Je me suis toujours demandé si c'était bien là la seule raison, car en fait, il faisait tellement chaud dans cette salle exposée plein sud, où l'on allumait becs Bunsen et bains-marie, qu'on n'y pouvait se mouvoir vraiment à l'aise qu'en très petite tenue! . . .

Les autres labos, répartis de part et d'autre du boyau central, étaient occupés par les chercheurs des groupes de François Jacob et Pierre Schaeffer. Dans le plus grand, Luiz Pereira da Silva, doux et amical, Harvey Eisen, agressif, et Jim Shapiro, plutôt débonnaire, formaient le trio des lambdologistes. Julian Davis, gallois épris de rugby, occupait la pièce dite des Américains, dont je devais hériter ultérieurement. C'est grâce à lui que j'ai dû de me retrouver à Colombes, quelques mois plus tard, pour le match France-Galles du Tournoi des Cinq Nations, isolé au sein du camp des supporters gallois! (Épisode qui se termina d'ailleurs fort joyeusement grâce à la bouteille de Johny Walker qu'il avait emportée et qu'il faisait circuler de bouche en bouche.) Un peu plus loin, on trouvait le groupe de Pierre Schaeffer, avec notamment Jeanine Michel, dont les convictions politiques résolument orientées devaient mener à bien des discussions, et Brigitte Cami, au caractère soupe au lait, qui mettait un peu d'animation, les jours de calme, par ses soudaines tempêtes de jurons.

Auprès de tant d'agitation, le groupe d'André était des plus calmes: on y percevait un feutré studieux, qu'entretenait sans effort l'autorité naturelle du "patron." En outre, Marguerite veillait.

Ce jour-là, je fus introduit dans ce que je considérais alors comme le saint des saints, la célèbre salle à manger sous la verrière en quoi se transformait tous les jours à 13 heures la pièce annexe de la bibliothèque de Jacques Monod. L'ambiance y était fort agréable et détendue. André et Marguerite se nourrissaient presque exclusivement de fruits, parfois d'un yaourt. André était très admiré d'ailleurs pour l'élégance qu'il montrait à peler poire, pêche, ou banane avec fourchette et couteau, et il n'hésitait pas à l'occasion à entreprendre l'éducation de la personne qui se trouvait à côté de lui. Bien peu, il est vrai, parvenaient à le suivre dans ce difficile chemin des bonnes manières.

Selon la coutume américaine, tant que j'étais aux États-Unis, je l'avais appelé par son prénom. Il ne me fallut pas longtemps pour apprendre que la traduction française du mot André, c'est: "Monsieur", et je devais provoquer une certaine stupeur, un jour à table, pour l'avoir oublié.

Nous eûmes ensuite une longue conversation scientifique, au cours de laquelle je pus lui présenter les résultats du travail que j'avais accompli avec David Baltimore au Salk Institute. Nous venions de mettre en évidence les intermédiaires de la réplication du RNA du poliovirus, qui renferment les molécules naissantes de RNA viral, appariées par leur extrémité en croissance à la matrice de RNA complémentaire. Nous avions caractérisé la structure cellulaire ribonucléoprotéique dans laquelle se retrouvaient ces intermédiaires, et nous lui avions donné le nom de complexe de réplication. Enfin, nous avions déterminé bon nombre des paramètres de la cinétique de réplication. Je ne savais trop, à vrai dire, comment lui exposer ces résultats assez techniques. Or, j'avais à peine commencé que déjà il anticipait, jugeait, critiquait, et concluait. Je fus frappé en outre, et cela se confirma à maintes reprises par la suite, par la soudaineté avec laquelle il pouvait faire fi de toutes ses hypothèses les plus chères, s'il se trouvait qu'un modèle expérimental nouveau les contredît. James Darnell m'avait appris, pendant deux ans, à travailler par une méthode en quelque sorte de "trial and error": face à un nouveau résultat, on recherche toutes les hypothèses possibles qui pourraient en rendre compte, puis on s'efforce de réaliser les expériences les plus élégantes, dont les résultats de tout ou rien permettront d'éliminer une à une celles des hypothèses qui sont erronées. Avec André, le processus intellectuel est tout différent. Servi par sa remarquable intuition, il choisit une piste, et décrète a priori que c'est la seule bonne. S'il pense qu'une hypothèse est fausse, il lui paraît inutile de s'y attacher, même pour en prouver la fausseté; il préfère, dans une vue plus synthétique, chercher à prouver la justesse de son modèle. Il aime citer ce mot, paraît-il d'Eddington: "Les seules expériences valables sont celles qui confirment la théorie." Et, ce qui déroute souvent ses interlocuteurs, il lui arrive de raisonner alors comme si la théorie était déjà démontrée! Mais il n'hésite pas un instant, ensuite, au vu de résultats expérimentaux probants, à renier son idée première et à s'enthousiasmer pour la suivante, si c'est nécessaire.

Septembre était là et le group de François Jacob n'avait pas encore déménagé, les travaux de réfection du rez-de-chaussée n'avançant guère. En outre, arrivait au laboratoire en même temps que moi Marc Fiszman, que le "patron" d'autorité baptisa Yves afin d'éviter les confusions auxquelles auraient pu prêter nos prénoms communs. André parvint néanmoins à me faire dégager un mètre cinquante de paillasse dans la toute petite pièce du fond du couloir. Je m'y installai tant bien que mal et entrepris d'y rédiger ma thèse. Il fit entreposer les caisses et cartons contenant mes cahiers d'expériences et dossiers divers dans son propre bureau, pour ainsi dire sous ses pieds. Je fus très touché de cette marque d'attention et de simplicité! Ces caisses contribuaient peu, évidemment, à améliorer l'aspect du bureau, naturellement austère et quelque peu vieillot. Mais la dignité d'André n'en souffrait nullement: je le trouvais, et bien d'autres comme moi, tout aussi intimidant qu'auparavant.

Je n'oublierai pas la grande patience qu'il déploya alors pour corriger mon manuscrit de thèse. L'aisance avec laquelle il manipule la langue française, l'aversion qu'il professe pour les anglicismes, en font un réviseur redoutable et un puriste redouté. "Il y a toujours un mot français pour tout dire—m'enseigna-t-il un jour—et s'il n'y en a pas, inventez-le!" Danièle Bucchini étudiait alors l'effet de divers dérivés de la guanidine sur le développement viral, et l'action antiguanidine des acides aminés et des dipeptides; il nous arrivait donc fréquemment d'employer entre nous le mot "tester": "Tester, Monsieur, cela veut dire faire son testament"! De même, l'usage abusif du verbe ramener pour le verbe rapporter provoquait son indignation: "Pour un objet, on dit: rapporter. La seule exception connue c'est: ramène ta fraise" nous dit-il un jour, excédé.

Sur ces entrefaites, il partit passer l'année dans le laboratoire de Renato Dulbecco à La Jolla. Pendant son absence, le poliovirus nous réserva à Marc Fiszman et à moi-même, quelques surprises: lors d'un passage à 40° au cours du cycle de multiplication, la synthèse du RNA viral s'arrêtait, et même la radioactivité préalablement incorporée disparaissait en majeure partie. Après plusieurs expériences de contrôle, force nous fut d'admettre que le transfert à haute température induisait une dégradation du RNA viral. Celle-ci était d'autant plus manifeste qu'on effectuait plus tardivement le changement de température, et la nucléase hypothétique responsable de cette dégradation préexistait toujours au transfert. Nous en vînmes donc à conclure que la température agissait par activation—ou démasquage—d'une ribonucléase.

André revint de Californie quelque six mois plus tard. L'hypothèse de la nucléase lui plut infiniment. Il en fit sa théorie. Mais une question se posa immédiatement: cette nucléase hypothétique était-elle d'origine virale? ou bien s'agissait-il d'une nucléase cellulaire? Les tenants des deux versions s'affrontèrent. Pour André, il s'agissait d'une modification, allostérique, de la réplicase virale, qui, perdant son activité polymérasique, acquérait à 40° une activité ribonucléasique. On savait que la DNA polymérase de Kornberg possède une activité désoxyribonucléasique intrinsèque; cela constituait donc un modèle applicable au cas qui nous intéressait. Mais, allant beaucoup plus loin, André imaginait qu'à l'origine, il n'y avait que des nucléases cellulaires, et qu'un jour l'une fut mutée, ce qui lui conféra une activité réplicasique jusque-là inconnue. Ainsi, le premier virus à RNA aurait eu pour génome le RNA messager muté qui spécifiait la séquence de cette ex-nucléase, laquelle maintenant pouvait le répliquer à volonté. Cette hypothèse nous stimula beaucoup, et à vrai dire, à ce jour encore, aucune preuve définitive n'est venue la réfuter. Néanmoins, nous ne pûmes mettre en évidence, dans les extraits de cellules infectées à 40°, aucune activité nucléasique particulière qu'on ne trouvât aussi dans les extraits de cellules témoins non infectées. Les préparations de réplicase virale dont nous disposions ne montrèrent aucune activité nucléasique à haute température. Toutes les souches virales que nous étudiâmes, quel que fût leur rt, provoquèrent toutes, à des degrés divers, le même phénomène de thermo-dégradation. Qui plus est, en relisant de récents articles de l'équipe d'André Kirn, de Strasbourg, nous découvrîmes qu'un phénomène analogue se produisait avec le virus vaccinal. Enfin, Jean-François Delagneau, à Grignon, faisait avec le virus de la fièvre aphteuse cultivé en cellules de hamster, les mêmes observations que nous avions faites avec le poliovirus en cellules de primates. Nous fûmes ainsi amenés à penser que la nucléase responsable de la thermodégradation pouvait être celle des lysosomes de la cellule infectée. Ceux-ci, on le sait, se fragilisent au cours du cycle de multiplication du poliovirus; ils finissent par éclater à la fin du cycle. Tout se passe donc comme si le passage à 40° entraînait un éclatement prématuré des lysosomes, avec pour conséquence l'arrêt de la multiplication du virus. La cellule infectée se suicide lorsqu'on la met à haute température, et, dans sa mort, elle entraîne le virus. Cette hypothèse, qui elle non plus n'a pas encore reçu de confirmation expérimentale définitive, finit par tant plaire au "patron" qu'il en devint le plus acharné zélateur.

Le départ de Pasteur et l'installation à Villejuif n'ont marqué de changements que dans les apparences. André assume désormais la lourde responsabilité de la direction d'un institut entier. Et pourtant, il n'était pas moins imposant dans la simplicité du décor pastorien que dans son vaste bureau directorial tout de cuir et de teck.

Il prépare courtoisement le café pour ses interlocuteurs de l'après-déjeuner. Sans se départir de son calme, il est parvenu à faire frémir la sacrosainte routine de la maison. A son habitude, il n'a pas hésité, chaque fois qu'il le jugeait nécessaire, à accepter des épreuves de force, voire à les provoquer. Il ne craint guère la contestation, car, lorsqu'elle se présente, il a l'habileté d'en saisir lui-même les rênes. Il doit, noblesse oblige, souffrir les affres de la télévision et des interviews et répondre que "non, on n'a pas encore trouvé le remède contre le cancer" mais que "oui, cela ne saurait tarder, à condition que la recherche puisse disposer de beaucoup de moyens humains et matériels" . . . Mais le Directeur d'Institut est resté Directeur du laboratoire. S'il a dû se décharger sur nous d'une partie de ses responsabilités, il suit néanmoins notre travail avec autant d'attention qu'auparavant, écoute, suggère, conseille et critique. Après m'avoir enseigné mon métier de chercheur, il s'attache désormais patiemment à m'apprendre celui de responsable scientifique.

Are/were mitochondria and chloroplasts microorganisms?

❦

SEYMOUR S. COHEN

Upon analysis all Insects generally yield some acid. In that they differ from other animals, and can be considered the transition from the règne animal *to the* règne végétal *which abounds in acid; just as cruciferous plants which give a great deal of volatile alkali, or gramineous plants which contain a very large quantity of mucous matter and can be considered the transition from the* règne végétal *to the* règne animal.

GUILLAUME-FRANÇOIS ROUELLE, 1762

FOREWORD

At one time the search for order, for connectedness, was a normal phenomenon in the development of scholarship and in the growth of a scientist. Such a habit of mind assumes a breadth of interest that reaches out to a world somewhat greater than that under experimental scrutiny. In this spirit Rouelle, the great teacher of French chemists, obviously believed that chemical phenomena might yield important clues to the structure of the biological world. André Lwoff has sought such clues at the cellular level for all of his career. At one critical stage in my own development he introduced me to the complexities of the cellular world in such a way that I began to perceive my own experimental work within the context of this larger world. Today one is thought to be wasting his time if he reads beyond the purview of his immediate experimental work, but I have always been grateful to teachers, such as André, who in their own scholarship have encouraged me to expand my outlook on the larger world and thereby to deepen my view of my own work.

The symposium at Cold Spring Harbor in the summer of 1946 was a landmark in the history of biology. Not only did this event present several important new discoveries in microbial genetics, for example, the findings of Lederberg and Tatum, Hershey, Delbrück, and Bailey, but also the meeting represented the emergence of a major new scientific community immediately after World War II. The meeting itself, as well as the whole summer of 1946 at Cold Spring Harbor, in fact had the flavor of a huge party, not only anticipating a great future for work in this field, but also releasing an enormous burst of energy and emotion after emergence from the pall of the war.

That summer was both a scientific and a personal landmark for me. The symposium was a revelation, introducing me to microbial physiology and genetics, subjects quite esoteric to someone trained in structural and analytical biochemistry. Although I had worked on the composition and structure of plant viruses and rickettsiae, I had begun the biochemical study of virus-infected bacteria a year earlier, in spite of knowing very little microbiology. In fact, I had come to Cold Spring Harbor hoping to fill in some of this deficiency by taking the phage course under Max Delbrück, my fellow students including Mark Adams, Earl Evans, and Birgit Vennesland. This course under Delbrück was, as is well known, a remarkable introduction to quantitative biology. Many of the symposium participants from Europe stayed on at the laboratory for the summer and even participated occasionally in the activities of the course. During that period I came to know the French contingent, André Lwoff, Jacques Monod, Boris Ephrussi, and Raymond Latarjet. The camaraderie and the scientific excitement of that summer were infectious. I realized very early that I had to learn a good deal of microbial physiology, and when I was invited to work in the Service de Physiologie Microbienne at the Institut Pasteur in Paris, the rather startling idea of a sabbatical year in recuperating Europe became an exciting reality. There is no doubt at all that the summer of 1946 gave me a clarity about my work that facilitated and sharpened my efforts in the ensuing productive year. It also revealed to me the social and intellectual potentialities of the life of a scientist in peacetime, a world that I had never suspected to exist.

In September 1947, my wife, my small son, and I went to Paris. André met us and drove us (through almost empty streets) to our hotel just off Boulevard St. Michel. The next morning the very first person I saw on the Boulevard was Jacques Monod, on his way to visit Boris Ephrussi a few streets away at the Institut de Biologie Physico-Chimique. I went with him to the Institut, and that morning I met for the first time the graduate student Piotr Slonimski, who had just begun his studies on the respiration of yeast mutants. As is well known, the discoveries of Ephrussi and Slonimski on "petite" mutations in 1949 became an important part of the data pointing to the genetic independence of mitochondria.

Once settled at the Service de Physiologie Microbienne at the Institut Pasteur, I began to collect André's papers. Although I had known that he was an acknowledged expert on microbial nutrition, I had not yet appreciated that he had begun his career as a protozoologist and had made numerous contributions to the physiology and nutrition of trypanosomes and *Hemophilus,* to the development of protozoan organelles and the survival of *Euglena* chloroplasts, etc. And I discovered his book published during the war, *L'évolution physiologique,* which introduced me to what has become a lifetime interest, the exploration of the biochemical origins of cells and their metabolic systems. Indeed I have chosen to teach in the area of biochemical evolution because I believe that the incorporation of the concept of evolutionary time into the thought and interest of students of biochemistry can enrich their work, even as I feel that my exposure to the work and thought of André Lwoff has enriched my own.

THE FITNESS OF THE SCIENTIFIC ENVIRONMENT

The question we have asked about mitochondria and chloroplasts—that is, are they symbionts within the cells they inhabit?—is old, having been posed even in the nineteenth century. The sunflares of biological and biochemical knowledge during the past quarter of a century have revealed numerous new paths of approach and investigation to this problem. The relatively recent explosion of activity of work on the biogenesis of these organelles has within 5 years resulted in hundreds of significant papers. The specific events which signaled the inception of this activity were the discovery of DNA in chloroplasts by Ris and Plaut in 1962 (1) and that of DNA in mitochondria by Nass and Nass in 1963 (2).

Of course these seminal observations were made after genetic evidence concerning the non-Mendelian inheritance of the organelles had been acquired over many years; this evidence is still being collected. Indeed the concept of the existence of cytoplasmic units endowed with genetic continuity has been actively promulgated throughout the past 25 years, of which Lwoff's analysis of the role of kinetosomes in ciliate development is one leading example (3). Important experiments such as that of Luck with choline-labeled mitochondria in 1963 (4) strongly suggested organelle division, that is, "existing mitochondrial material is directly and randomly transmitted to progeny in formation of new mitochondria." It may be noted that Luck's paper was written before the existence of mitochondrial DNA was well known. Furthermore, a well-developed literature, less familiar perhaps to biochemists and molecular biologists, existed on the intracellular symbiosis and parasitism (these terms may frequently be difficult to distinguish by experimental or ecological criteria, as pointed out by Theobald Smith) of bacteria and blue-green algae. Blue-green algae frequently inhabit eucaryotic cells (5), and chloroplasts may function within animal cells in nature (6).[1] Nevertheless the actual demonstrations that these organelles contained the genetic material, DNA, appears to have aroused a scientific public for whom these three magic letters had suddenly (over a decade) assumed the role of an addictive mind-expander.

At a time when the role of DNA in the cellular division of labor was emerging in essentially unchallenged pre-eminence, it was only natural that the problems of polymer synthesis dominated by DNA should have drawn first attention, pointing to even more specific bio-

[1] My colleague, M. Nass, has recently fed chloroplasts to mouse fibroblasts. The resulting enlarged green cells, each containing many chloroplasts, can still grow and divide. The chloroplasts may be re-isolated after several days and have been shown to retain their DNA and photosynthetic activity (7).

logical questions. These questions were, "Does the DNA of the organelle act as the genetic substance of these organelles, and is this substance different from the nuclear DNA of the eucaryotic cells in which they operate?" and "To what extent do these genetic systems of the organelle determine their structure and function?"

As one might expect, these specific questions have not yet been answered in their entirety, but the general question of our title is clearly not trivial. The problem may now be posed as a serious scientific challenge, warranting the most systematic and penetrating exploration. After all, if in the history of science we honor the Copernican revolution as the demonstration that man is not the center of the Universe, what effort should be accorded the proposition that man (and indeed all higher organisms) may be merely a social entity, combining within his cells the shared genetic equipment and cooperative metabolic systems of several evolutionary paths? We suspect that governments should be interested in such a possibility, but they may not yet have heard about it, nor might their responses be readily predictable.

ON THE BIOCHEMISTRY OF GENETIC INDEPENDENCE

The specific questions posed above have been tested by all the modern and growing armamentarium of molecular biology. The DNA's of mitochondria and chloroplasts have been isolated and characterized with respect to size and shape, density, and base composition. A mitochondrion appears to contain multiple (2–6) copies of its DNA, each in individual compartments, as in some filamentous bacteria which have managed to replicate but not to divide into separate cells. The DNA of animal mitochondria, for example, is double-stranded and circular, but it is small (only about 5 μ long or about 10×10^6 molecular weight) (8) and therefore unable to code for all the proteins and enzymatic activities found in a mitochondrion. Nevertheless the mitochondrial DNA differs from the nuclear DNA in base composition, and the two DNA's do not interact in hybridization experiments. The DNA's of plant and fungal mitochondria also appear clearly different in polynucleotide sequence from those of their respective nuclear DNA's.

The mitochondrial DNA in plants, although not as well studied as the animal material, shows some interesting differences from that in animals, being isolated as significantly longer linear pieces, perhaps more nearly approaching the DNA found in chloroplasts. Indeed chloroplast DNA does appear to exist in a size comparable to that of a bacterium or a blue-green algae, that is, 10^{-15} to 10^{-14} g per chloroplast, depending on the source (9, 10), and therefore chloroplast DNA probably codes for many more proteins than does mitochondrial DNA.

The organelles are equipped with apparently characteristic enzymes active on DNA, that is, both DNA polymerase and DNA-dependent RNA polymerase, which are presumably active in DNA replication and transcription *in situ*. They also appear to contain the entire panoply of metabolic equipment necessary to make specific proteins.

The products of transcription found in the organelles include several classes of RNA, such as two characteristic ribosomal RNA molecules and at least several species of transfer RNA. Little is known of other classes of RNA, such as messenger RNA, in the organelle. The mitochondrial RNA fractions categorized as ribosomal are composed of two nucleic acids thought to derive from the mitochondrial ribosomal subunits. These nucleic acids, as well as the mitochondrial ribosomes and their subunits, possess slightly but significantly smaller sedimentation rate constants than do their cytoplasmic counterparts. Similar findings have been obtained for these components of spinach and *Euglena* chloroplasts, as compared to the cytoplasmic constituents. In these respects, the mitochondrial and chloroplast structures resemble bacteria, whose ribosomes and rRNA are also significantly smaller than the comparable units of eucaryotic cytoplasm. However, these components of the organelles and of bacteria do differ somewhat among themselves, and as yet the data do not permit any designation of phylogenetic relationships. Furthermore the hybridization data in some instances have indicated a small cross reaction of mitochondrial and chloroplast RNA with their respective nuclear DNA, suggesting either that some organelle RNA may be made in the nucleus, or that the RNA isolated from the organelles may be slightly contaminated with cytoplasmic components.

The study of other structures involved in translation has produced clearer evidence of intramitochondrial biogenesis. Mitochondrial amino acid acyl synthetases were shown to differ from the cytoplasmic enzymes. Transfer factors derived from either *Escherichia coli* or *Neurospora* mitochondria were found to be relatively active for bacterial and mitochondrial ribosomes and far less active with the ribosomes of *Neurospora* cytoplasm. Thus bacterial and mitochondrial factors showed marked functional homology (11), and it was also shown that transfer factors from *Neurospora* cytoplasm were inactive with *E. coli* ribosomes and much more active with cytoplasmic than with mitochondrial ribosomes.

Even more definitively, mitochondrial tRNA's are found not only to differ from their cytoplasmic counterparts, but also to hybridize specifically with mitochondrial DNA (12, 13).

The initiation of protein synthesis in bacteria, mitochondria, and chloroplasts seems to have common features, involving the formation of *N*-formylmethioninyl–RNA and the beginning of the polypeptide chain with *N*-formylmethionine. This process, common to bacteria and these organelles, clearly differs from the inception of protein synthesis in the cytoplasm of eucaryotic cells.

Thus, in a very few years, molecular biology has revealed astonishingly close similarities among procaryotic cells and the DNA-containing cytoplasmic organelles of eucaryotic cells in the systems active in the synthesis of RNA and proteins. Of course numerous major points as well as fine details have still to be clarified. For example, it will be important to compare the elements of transcription and translation among the nucleus and cytoplasm, mitochondria, and chloroplasts of a single type of plant cell to see whether the three types of genetic system are all structurally different in these cells. As will be seen below, a perfectly reasonable question is whether the ribosomal proteins of cytoplasm, mitochondria, and chloroplasts have some units in common. It would interest me to know whether polynucleotide phosphorylase, an enzyme possibly involved in mRNA degradation, which is unequivocally present in bacteria and blue-green algae (14) and has been reported to be present at a low concentration in spinach (15), is actually present in chloroplasts or mitochondria or in both.

MITOCHONDRIACS AND ORGANELLE COMPOSITION

The long pursuit of the mechanisms of respiration and of oxidative phosphorylation in eucaryotic cells has led to a chemical and structural dissection of the mitochondrion. After some years of fractionation and structural analysis, we may speak of chemically distinguishable and physically separable outer and inner membranes. From the latter the internal invaginated fingers of membrane, the cristae, as well as DNA circles, extend into a central matrix. The outer and inner membranes are rich in lipid; nevertheless removal of all lipid does not appear to destroy the double-layered structure of the inner mitochondrial membranes and cristae.

The lipids of mitochondria include some substances not commonly found in bacteria, such as the choline-containing lecithin and sphingomyelin; these are predominantly in the outer structure. Although the organelle does not contain all the enzymes for the assembly of complex phospholipids, the outer membrane appears to possess transferases which permit numerous steps in fatty acid exchange, and choline incorporation into lecithin. Acetate is also incorporated into fatty acid in isolated mitochondria, but such a process appears to involve mainly an extension of preformed fatty acid, as in the conversion of palmitate to stearate. This mitochondrial activity does not involve the formation of malonyl derivatives. Nor can the mitochondria desaturate the long-chain fatty acid stearate to oleate. This reaction, which is characteristic of cytoplasmic synthesis, is similarly missing in bacteria and mitochondria.

The organelle is rich in the phosphatidyl glycerol derivative cardiolipin, which is not known in the remainder of the cell; this substance is characteristically found rather widely in bacteria. Similarly mitochondria are quite low in sterols (bacteria are usually devoid of sterols unless they are *Mycoplasma* grown in the presence of sterols). In addition, it is of interest that polyunsaturated fatty acids (linoleate and γ-linolenate), also absent from bacteria, are reported to be nonessential to mitochondrial development, structure, and function in some special fibroblast strains after 10 years of subculture in lipid-free media (16).

Thus the so-called essential fatty acids do not appear essential to respiration and oxidative phosphorylation, at least in these cells.

A similar situation has been found with respect to chloroplast structure and photosynthesis. Organisms capable of photosynthetic O_2 production, from the blue-green algae on up, were found in initial analyses to possess characteristic lipids, including such substances as galactosyl diglycerides and an unusual sulfolipid, both of which contained large amounts of the polyunsaturated acid α-linolenate.[2] Characteristically the anaerobic photosynthetic bacteria, like most bacteria, lack this fatty acid and contain as unsaturated fatty acid mainly monoenoic acid, for example, cis-vaccenic acid. This distribution seemed to suggest a role for α-linolenate in O_2 evolution. However, the O_2-evolving blue-green alga, Anacystis nidulans, was found to lack this fatty acid (17). In hindsight, it is difficult to understand how α-linolenate, presumed to be derived from oleate, which itself is derived from stearate via an O_2-requiring reaction, could be imagined to be essential to O_2 production. Nevertheless I suspect that eventually we shall learn that α-linolenate, in some subtle or not so subtle way, does help in the development of a conformation or environment favorable to O_2 production, just as γ-linolenate and other essential fatty acids requiring O_2 for their biosynthesis, in a later evolutionary step, may have facilitated the production of complexes in which certain respiratory complexes and the machinery of oxidative phosphorylation can operate stably and optimally.

Mitochondrial membranes contain numerous proteins, among which are found many enzyme activities characteristic of the organelle. The integrated elements of the respiratory chain are found mainly in the inner membrane. From these basal structures arise the characteristic stalked spheres which contain the oligomycin-sensitive ATPase, apparently essential to oxidative phosphorylation. These structures and sequences are the core of the matter, providing the crucial steps by which the small discrete steps of the free-energy change of electron transfer, terminating in the acceptor, O_2, are transduced to form ATP.

An eucaryotic cell deficient in this mitochondrial activity will be-

---

[2] It is of interest that blue-green algae and leaf chloroplasts do not contain phosphatidyl choline, phosphatidyl ethanolamine, phosphatidyl inositol, and cardiolipin, although these are present in leaf extracts (perhaps in leaf mitochondria and other structures) (17).

have in most aspects like an anaerobe and will grow slowly in the presence or absence of O_2. Nevertheless it is important to remember that such a cell defective in mitochondrial function may still possess many O_2-utilizing reactions. With only a few exceptions these are centered in the lipid-rich membranes of the endoplasmic reticulum, and control such essential functions as sterol biosynthesis, the formation of oleate, the conversion of gulonolactone to ascorbate, the hydroxylation of phenylalanine to tyrosine, and the oxidation of tryptophan to M-formylkynurenine (and through this to nicotinic acid). These O_2-requiring reactions are characteristic of the eucaryotic cell and to a significant extent are essential to the profusion of membranous structures found in such cells. Thus anaerobic yeast requires an exogenous sterol for growth.

In almost every instance such functions are determined by nuclear genes. It appears therefore that, although the division of labor in a eucaryotic cell has centered the respiratory function in bodies which bear some resemblance to bacteria, numerous acquisitions of enzymatic function crucial to the structure of eucaryotic cells, for example, lipophilic oxygenases and other O_2-utilizing systems, are actually determined genetically and constructed outside the mitochondrion.

These "new" activities are almost totally absent from all but a very few bacteria, such as some pseudomonads. The aerobic bacteria are thought to have arrived relatively recently on the phylogenetic scene. Some of these rare bacterial oxidative enzymes, such as phenylalanine hydroxylase, tryptophan pyrrolase, and cytochrome oxidase, have actually been purified, even crystallized, from pseudomonads. Although the mechanisms of the bacterial and mammalian enzyme (e.g., phenylalanine hydroxylase) may have numerous similarities, the bacterial enzymes and eucaryotic enzymes do have many differences. Nevertheless the examinations have not proceeded in these instances to the level of polypeptide sequence, that is, to the search for homology, to determine whether there is in fact any genetic relation between these procaryotic and eucaryotic enzymes, as has been found among the cytochrome c's of eucaryotic cells.

When we return to mitochondrial enzymes, many other significant opportunities for the detection of homology become evident. For example, we might ask about the possible homology among components in α-keto acid (pyruvate or α-ketoglutarate) dehydrogenase complexes.

The classical studies of Reed and his collaborators (18) have revealed remarkable similarities between the complexes derived from *E. coli* and mitochondria with respect to details of size, structural features, and mechanism. Can these similarities represent homology or an extraordinary evolutionary convergence, that is, an independent discovery of optimally efficient structure? One can hardly wait for a chemically clear answer.

It would be unfair to leave this sketchy summary without recording a curious and potentially disturbing anomaly. It is well known that bacteria, a few fungi, and almost all plants (other than euglenids) make lysine exclusively via α,ϵ-diaminopimelic acid (DAP), which is a characteristic component of many bacterial cell walls (as well as those of blue-green algae[3]). The formation of certain intermediates in the biosynthesis of DAP appears also to be essential to sporulation, another unique bacterial activity. On the other hand, fungi such as yeast make lysine via a series of reactions in which homocitrate is converted to α-aminoadipate (AAA), eventually to saccharopine [ϵ-N(L-glutaryl-2)-L-lysine], and finally to lysine. Many of these reactions, such as the formation of homocitrate, occur in the mitochondria. Of course an animal is incapable of synthesizing lysine, but the reason is not quite clear, since lysine is metabolized by animal mitochondria, in an apparent reversal of the biosynthetic steps, to saccharopine and other components of the AAA pathway. Thus animal mitochondria appear to contain remnants (?) of a path of lysine biosynthesis absent from bacteria and found in some fungal mitochondria.

WHERE DO THE PROTEINS COME FROM?

The complete machinery for protein synthesis, from DNA template on, seems to be present in the organelle, but the information potentially contained in this DNA is too limited to code for all the proteins and enzymes found in the mitochondria. Assuming the homogeneity of

[3] Do chloroplasts contain DAP, and is all lysine production in higher plants a chloroplast function? No one has yet tested this question, nor has the apparent anomaly of euglenid lysine synthesis via α-aminoadipate been explored in appropriate depth.

mitochondrial DNA, it has been estimated that at least 30% of this material of 10×10^6 codes for species of mitochondrial rRNA and tRNA. Thus we are left with information for at most about 18 mitochondrial proteins of molecular weight 20,000. Which of the very many proteins and enzymes found in the organelle are actually coded for and made in the mitochondria?

The "petite" mutants of yeast are characterized by damaged mitochondria and are unable to respire. Such mutants, induced by dyes such as acriflavin or ethidium bromide, display the now classical pattern of cytoplasmic inheritance. The loss of a protein component from "petite" mitochondria can be seen in polyacrylamide gel electrophoresis (19). More specifically the mutant mitochondria have been shown to lack an inner membrane component essential for the attachment of otherwise apparently normal ATPase to the inner mitochondrial structure (20). It is of interest that the loss of this component does not merely affect oxidative phosphorylation, some components of which are controlled by nuclear genes (21). "Petites" contain cytochrome c but lack a number of heme components, for example, cytochromes b and $a(a_3)$, and do not possess a functional cytochrome oxidase. It has been found that the apoprotein of cytochrome oxidase is present in the mitochondria of these mutants and that a functional cytochrome oxidase can be reconstituted by incubation of mutant submitochondrial particles with heme a^4 (22). The conclusion was thus reached that in this important class

---

[4] The cytochrome oxidase of eucaryotic cells contains copper, which is believed to be essential to the enzymatic activity. However, the appropriate fractions of the "petite" mitochondrial extract contained normal amounts of copper (21) consistent with the block occurring specifically in heme a biosynthesis. Substances that chelate copper, such as cuprizone, produce giant, deformed mitochondria, and a deficiency in copper transport, as in Wilson's disease in the mammal, leads to a deficiency of cytochrome oxidase. Copper is also present in other enzymes of eucaryotic cells, for example, the monoamine oxidase of mitochondria or plastocyanin and ribulose diphosphate carboxylase of chloroplasts. However, the cytochrome oxidase (a_3) of *Pseudomonas* lacks copper. Indeed it has not been proved that any procaryotic cell has functional copper. Although the *Pseudomonas* possess a copper protein capable of reducing cytochrome oxidase *in vitro,* it has not been demonstrated that this protein has any role in the respiration of the organism. The presence of copper in the plastocyanin and ribulose diphosphate carboxylase of chloroplasts poses the problem of the possible existence of this metal in blue-green algae and in some of their presumably colorless descendants, such as the flexibacteria and some autotrophs.

of mutants the mitochondria had lost an important step in the biosynthesis of heme a which was coded for and synthesized by the organelle. It may be noted that the cytoplasmic mutant of *Neurospora* (poky), which also have respiratory-deficient mitochondria, possesses a defective cytochrome oxidase in which the protein itself is altered (23).

Protein synthesis in isolated mitochondria as in bacteria, but unlike that of cytoplasmic ribosomes, is quite sensitive to chloramphenicol. On the other hand protein synthesis in cytoplasm is very sensitive to cycloheximide, unlike the bacterial and mitochondrial syntheses. It may be mentioned that the chloroplast pattern of sensitivity also resembles the mitochondrial or bacterial pattern. Amino acid incorporation into isolated mitochondria normally labels some components of the inner membrane, and it appears that under such conditions one major fraction of these components is inhibited by low levels of chloramphenicol (24). In whole cells the antibiotic prevents the synthesis of several internal mitochondrial enzymes, including cytochrome oxidase, without inhibition of the synthesis of cytochrome c (25). However, at high concentrations the antibiotic rapidly inhibits respiration and cell growth (26), as well as mitochondriogenesis (25).

Inhibitor techniques have been used to show not only that enzymes such as cytochrome oxidase or those involved in its biogenesis are mitochondrial in origin but also that numerous others, such as the proteins of the mitochondrial ribosomes, are actually synthesized on cytoplasmic ribosomes (27). It may be mentioned that kinetic techniques used to show that cytochrome c of the mitochondria is derived from cytoplasmic synthesis (28) have also been interpreted as indicating that the protein of cytochrome oxidase is probably not constructed in the mitochondria. This conclusion seems to be at variance with the result obtained with *Neurospora* (poky).

Anaerobically grown yeast contain poorly formed mitochondria (promitochondria) which contain DNA and incomplete inner membranes (29). Sterols are not essential to such anaerobic syntheses. The membranes of the promitochondria lack cyanide-sensitive respiration and numerous cytochrome continuants but can adapt to form these components on aeration in low-glucose media (30). The nature of this oxygen-induced adaptation has been quite puzzling for many years. Most recently it has been found that a burst of mitochondrial DNA synthesis precedes the formation of respiratory activity (31). Both cycloheximide and chloramphenicol block the O_2-induced adaptation, although cycloheximide did not inhibit synthesis of mitochondrial DNA (32). Thus analysis of this remarkable response to the gaseous environment has similarly revealed the dependence of mitochondrial biogenesis upon both its own genetic and metabolic apparatus and other nuclearly determined metabolic systems.

Careful electrophoretic analysis of the membranes of the mitochondria has revealed at least 12 and 23 components in the outer and the inner membranes, respectively. These membranes appear to have only a single common protein between them (33). Of the 15 proteins detected by this technique in the microsomal membrane, only about 3 seem to be present among components of the outer membrane. Despite the apparent segregation of these proteins in the mitochondrial compartment, there appears to be a growing agreement that most of the mitochondrial proteins are of cytoplasmic origin. The problems of the mechanisms by which nuclear proteins made in the cytoplasm reach their specific sites in the nucleus can now be extended to cytoplasmic organelles as well.

ON THE COMPARATIVE BIOCHEMISTRY
OF O_2-PRODUCING ORGANISMS

At first sight it might seem a bit easier to relate chloroplasts to bluegreen algae. Cytoplasmic inheritance of the organelle is well known. Chloroplasts have a larger complement of DNA, presumably containing a greater number of polynucleotide sequences supposedly translatable into a large number of characteristic proteins. The presence of polymerases, ribosomes, tRNA, etc., has been affirmed, and indeed chloroplast systems *in vitro* do very well in making complete proteins, such as tobacco mosaic virus subunits. Blue-green algae fulfill the role of photosynthetic symbionts in many systems, and ingested plants can furnish symbiotic photosynthetic chloroplasts even in animal tissues. The observable differences in the photosynthetic membrane structures, from the vesicles and lamellae of photosynthetic bacteria to blue-green algae to the fused membranes of spinach chloroplasts, are falling into place as a continuous sequence of steps in the evolution of invaginations from the inner cytoplasmic membrane of the bacterium, alga, and organelle.

The blue-green algae and eucaryotic chloroplasts both function with chlorophyll a instead of bacteriochlorophyll. The former pigment is metabolically more primitive, by the law of Granick. This fact, in addition to the peculiarities and distribution of photoassimilation and photosynthesis by the blue-green algae, has suggested that the common ancestor of both groups carried out photoassimilation with the aid of chlorophyll a. The evolution of bacteriochlorophyll systems in facilitating light gathering reduced the possible free-energy change and prevented the potential development of O_2 evolution, for which the presence of a pigment such as chlorophyll a is essential. Blue-green algae (and chloroplasts), unlike photosynthetic bacteria, thus have been able to evolve the linked, elaborate photosystems I and II, which culminate in the Hill reaction and O_2 evolution on one side (of the potential diagram) and reduced ferredoxin and NADPH at the other.

The O_2 thus evolved has permitted the biosynthesis of sterols, oleate, ascorbate, etc., as well as the formation of the "bile pigment" prosthetic group of the phycobilins, which help the blue-green algae (and the eucaryotic red algae as well) to gather more light to effect more photosynthesis.[5] Blue-green algae and chloroplasts have also evolved similar enzyme systems crucial to their specialized photosynthetic CO_2 fixation. These include the characteristic ribulose diphosphate carboxylase, which generates phosphoglyceric acid and is a larger and more highly organized enzyme in blue-green algae and in chloroplasts than in most photosynthetic bacteria.

In photosynthetic bacteria, such as *Chromatium,* both the photosynthetic cycle and glycolysis function on an NAD-utilizing triose phosphate dehydrogenase. A triose phosphate dehydrogenase specific for NADP is found only in eucaryotic chloroplasts and in O_2-evolving photosynthetic microorganisms. The fact that the appearance of this enzyme

[5] It would be of interest to know whether O_2 is also essential in blue-green algae and chloroplasts for the conversion of coproporphyrinogen to protoporphyrin IX, as it apparently is in mitochondria. Isolated chloroplasts are capable of converting δ-aminolevulinate to heme, but it is not yet known whether the formation of protoporphyrin in the chloroplast is an evolutionary remnant comparable to protoporphyrin synthesis in an obligate anaerobe, such as *Desulfavibrio.*

parallels the rate of chlorophyll synthesis in regreening *Chlorella* has suggested that the NADP enzyme is active specifically in the photosynthetic carbon cycle, while the well-known NAD enzyme in these cells acts primarily in glycolysis and oxidative reactions.[6]

Despite these differences between photosynthetic bacteria and chloroplasts, it is exciting to record the apparent existence of structural homology with respect to amino acid sequences among ferredoxins of bacteria and chloroplasts (34). Such studies may eventually serve to demonstrate an evolutionary sequence from procaryote to eucaryote organelle. However, no data are yet available on such a sequence in the ferredoxins of the presumed intermediate, the blue-green algae.

ON THE BIOGENESIS OF CHLOROPLASTS

An examination of the biosynthesis of chloroplast components has revealed a complex division of labor every bit as puzzling as that we have seen in our examination of mitochondrial independence. Despite the existence of non-Mendelian inheritance governing chloroplast prop-

---

[6] Some recent findings in the field of nitrogen fixation are so interesting in an evolutionary sense that I feel compelled to include them, even though they may seem a bit off the main subject of our discussion. In the last two decades nitrogen fixation has been revealed as a reductive capability of many organisms possessing a bewildering assortment of oxidative activities. These range from aerobes such as *Azotobacter* and the rhizobia, which symbiotically infect legume root modules, to anaerobes such as the clostridia and the photosynthetic bacteria. Some filamentous blue-green algae also fix N_2 but green leaves do not, and this is one more significant difference between blue-green algae and chloroplasts. The study of N_2 fixation in extracts has revealed the importance of the exclusion of O_2 from the reductive enzymes, even in such systems as the "aerobic" symbiotic rhizobia or the blue-green algae. Recently N_2 fixation in filamentous blue-green algae has been revealed as the activity of the unusual and previously mysterious heterocyst cells. Such cells have reorganized their lamellae for this reductive function and have lost the capacities of CO_2 fixation and presumably of O_2 evolution (33). It appears to have been demonstrated that the symbiotic N_2-fixing blue-green algae in the root nodules of Cycadaceae perform this function with a poor rate of CO_2 fixation. Thus the specialized N_2-fixing cells of the blue-green algae seem to have taken a step back to life as a bacterium. The O_2-evolving chloroplasts have learned to live with the consequences of this evolution of their membranes and enzymes, which have included a greater dependence on other organisms for the reduction of N_2.

erties, the genetic analysis of steps in the synthesis of chloroplast constituents has demonstrated that many controlling genes are nuclear. These
determine such functions as the synthesis of carotenoids, chlorophyll a,
and plastoquinone. In studies of *Chlamydomonas,* it has been found
that all mutations affecting photosynthetic electron transport are mutations in nuclear genes (35). In selected mutants of this organism, growth
in the dark leads to cells which lack chlorophyll and membranes and
discs containing this pigment. Greening occurs in such cells in the light,
as does photoreductive power. Chloramphenicol does not significantly
block greening and membrane formation, but does block the fusion of
membranes to form grana and the development of photoreductive activities characteristic of the two photosystems. On the other hand, cycloheximide blocks greening functions without affecting the proportionate
development of photoreduction (36). In other studies (37) it has been
observed that chloramphenicol also blocks the synthesis of ribulose diphosphate carboxylase and the NADP-specific triose phosphate dehydrogenase, whereas cycloheximide does not. The syntheses of these enzymes
are thus thought to center in the chloroplast, in contrast to those relating to electron transport.

The totality of evidence, of which that presented above is but a
small part, clearly indicates that, as in the case of mitochondrial biosynthesis, the synthesis of chloroplast structure and functional entities
involves the products of both nuclear and chloroplast genes and the
activity of both cytoplasmic and chloroplast ribosomes. Thus, for both
of these organelles, inquiry as to their evolutionary origins and their
degrees of genetic, biosynthetic, metabolic, and functional independence
has now revealed a previously unsuspected degree of complexity. The
organelles have established an elaborate interplay with the nucleus and
cytoplasm of their respective cells such that both sets of genomes and
ribosomes are essential to organelle biosynthesis and function. In all
eucaryotic cells the close integration of both sets of these structures and
their metabolic activities is crucial to the development of a successful
life cycle of the cell as a whole. Nevertheless, despite the necessary addition of new, more sophisticated questions to our inquiry, the possibility
that chloroplasts may have stemmed more or less directly from blue-
green algae symbionts is still very real. This inquiry evidently requires
many detailed searches for homology among both their polynucleotides

and their polypeptides, searches which have barely begun. Are the evidences for such descent—for example, the apparent homologies of bacterial and plant ferredoxin or those of 5S RNA of *E. coli* ribosomes and of KB cell ribosomes (38)—real, or are they illusory? These questions will certainly be resolved in this coming decade.

Furthermore, how shall we explore the possibility, for which we have no example at the present time, that the symbiotic procaryote, ancestral to the organelles, may have lost a large portion of its genome to become bits of the nuclear genetic complement? Will some DNA viruses prove to represent terminal events of such an evolutionary process?

ON THE ORIGIN OF EUCARYOTIC CELLS

Even if the answers had been simple and affirmative (i.e., yes, a mitochondrion is the direct descendant of this or that ancestral *Pseudomonas, Rickettsia,* or *Mycoplasma,* or a chloroplast is the direct descendant of this or that blue-green alga), the inquiry inevitably would pose a far more difficult question. What is the nature of the cell with which the presumed symbiont established its early working relation? It is possible that the inherent difficulty of this question forestalled serious inquiry until the discovery of DNA in the organelles made the problem inescapable.

Until very recently the theories of cellular evolution have assumed a more or less continuous development of eucaryotic cells from procaryotes. Such evolutionary and phylogenetic schema have largely tended to minimize the sharp discontinuities in the geological, cytological, compositional, and metabolic records which relate specifically to the differences between such cells. In the main these are differences which have appeared with the production of atmospheric O_2 and of blue-green algae. If such discontinuities had been widely recognized and had become a matter for consideration, it is possible that the classical phylogeneticists would have constructed hypotheses relating the development of the cytoplasm of a eucaryotic cell to the structural and functional differentiation of a procaryotic membrane. If we place almost all of the membrane-forming O_2-utilizing systems in the endoplasmic reticulum outside of the mitochondria, it can be imagined that the acquisition of a copper-containing cytochrome oxidase might have evolved as a late step in the evolution of the membrane, which formed a cytoplasmic nodule made of fatty acid and keto acid dehydrogenases and electron-transport and ATP-generating systems, plastered together with lipids and other membrane glues. A similarly unsubstantiated fantasy must eventually hold in regard to the production of the organelles of a primitive O_2-evolving eucaryote from the hypothetical ancestral blue-green alga. It is of interest that no one has attempted as yet to come to grips seriously with this type of hypothesis. Of course there would be many more problems to solve, even if these major theoretical steps were adjudged reasonable for such phylogenies.

Within these classical phylogenies, which have already begun to be rejected by serious biologists (39), the alternative concept of the existence of our organelles as once free-living symbionts creates other major problems concerning the possible nature of the eucaryotic host. Bluntly stated, can we imagine a viable eucaryote devoid of all the enzymatic apparatus, such as cyanide-sensitive respiration, present in the mitochondria?

The most serious effort in pursuit of the symbiont hypothesis to date has been that of Margulis (40, 41), who has attempted to construct a comprehensive picture of the evolution of all eucaryotic cells as the product of the symbiotic fusion of several procaryotic cells. In a most daring and serious effort, she has explored the geochemical and paleontological record, as well as the data of cytology, microbiology, and biochemistry, to seek the relevant information. It may be noted that much pertinent cytological and embryological evidence assiduously collected in the first third of this century has tended to be neglected by many classes of modern biologists; Margulis has also attempted to place this evidence in the context of the symbiotic theory. Briefly, it is postulated that eucaryotic cells arose between 0.5 and 1.0 billion years ago, perhaps a billion years after the evolution of O_2-forming blue-green algae. The presence of atmospheric O_2 presumably permitted the evolution of aerobic bacteria. The ancestral eucaryote then represents the fusion of an aerobic bacterium and another procaryote to give rise to a mitochondrion-containing amoeboid cell. The subsequent addition of a spirochaete-like organism provided the motile ancestral amoeba form, containing the crucial 9 + 2 flagellar symbiont subsequently active in the differentiation of centrioles and mitotic apparatus. This primitive cell may have evolved to form protozoa, fungi, or eventually multicellular animals. Another later possibility was "infection" of this cell by a blue-green alga to generate the ancestral eucaryotic alga on the way to the formation of higher plants.

At this point in my reading, this hypothesis seems to me no less likely than does any other. At the very least any existing symbiotic hypothesis on these questions is oversimplified, as I have detailed in the body of this essay. We must account not only for the multiplicity of discrete separated genomes in nucleus and organelles, but also for the mechanism of their fragmentation and assembly. We must also account for the division of labor and mechanisms by which a nuclear genome provides many gene products essential to the structures, function, and activity of cytoplasmic organelles that contain their own systems of genetic determination and polymer synthesis. However, even without pressing to this unexpected level of complexity, it will possibly take years or decades to evaluate Margulis' bare scheme, which in its most elaborate detail is just finding its way to a publisher. Will the details or the broad outline of this hypothesis hold up under the close, determined scrutiny which is clearly on its way in this decade?

I do not know whether or not André Lwoff will like the Margulis schema. It seems likely that his experience in classical cytology, cytoplasmic inheritance, and molecular biology makes him one of the very few scientists who can evaluate this hypothesis in a knowledgeable way, or at least tell us whether it warrants our serious consideration. In any case, I offer him the most recent data for the puzzle on which he has spent much of his scientific life. I assure him that his views on this matter would be a very useful guide to the biochemists who have at long last come to close grips with cytoplasmic heredity.

REFERENCES

1. H. Ris and W. Plaut, *J. Cell. Biol.*, 13, 383 (1962).
2. M. M. K. Nass and S. Nass, *J. Cell. Biol.*, 19, 593 (1963).
3. A. Lwoff, *Problems of Morphogenesis in Ciliates*, New York, John Wiley and Sons, 1950.

4. D. J. L. Luck, *Proc. Nat. Acad. Sci. U.S.*, 49, 233 (1963).
5. W. T. Hall and G. Clause, *J. Phycol.*, 3, 37 (1967).
6. D. L. Taylor, *J. Mar. Biol. Assoc. U.K.*, 48, 1 (1968).
7. M. M. K. Nass, *Science*, 165, 1128 (1969).
8. M. M. K. Nass, *Science*, 165, 25 (1969).
9. J. T. O. Kirk, *Biochemistry of Chloroplasts*, vol. I, p. 319, New York, Academic Press, 1966.
10. M. Edelman, D. Swinton, J. A. Schiff, H. T. Epstein, and B. Zeldin, *Bact. Rev.*, 31, 315 (1967).
11. H. Küntzel and H. Noll, *FEBS Letters*, 4, 140 (1969).
12. C. A. Buck and M. M. K. Nass, *J. Mol. Biol.*, 41, 67 (1969).
13. M. M. K. Nass and C. A. Buck, *Proc. Nat. Acad. Sci. U.S.*, 62, 506 (1969).
14. I. Capesius and G. Richter, *Z. Naturforsch.*, 22B, 204 (1967).
15. D. O. Brummond, M. Staehelin, and S. Ochoa, *J. Biol. Chem.*, 225, 835 (1957).
16. Y. Kagawa, T. Takaoka, and H. Katsuta, *J. Biochem. (Japan)*, 65, 799 (1969).
17. A. T. James and B. W. Nichols, *Nature*, 210, 372 (1966).
18. L. J. Reed and R. M. Oliver, *Brookhaven Symp.*, p. 397 (1968).
19. H. Tuppy, R. Swetly, and I. Wolff, *Europ. J. Biochem.*, 5, 339 (1968).
20. G. Schatz, *J. Biol. Chem.*, 243, 2192 (1968).
21. L. Kovac, T. M. Lachowicz, and P. P. Slonimski, *Science*, 158, 1564 (1967).
22. H. Tuppy and G. D. Berkmayer, *Europ. J. Biochem.*, 8, 237 (1969).
23. D. L. Edwards and D. O. Woodward, *FEBS Letters*, 4, 193 (1969).
24. S. Yang and R. S. Criddle, *Biochem. Biophys. Res. Comm.*, 35, 429 (1969).
25. A. M. Kroon and H. DeVries, *FEBS Letters*, 3, 208 (1969).
26. F. C. Firken and A. W. Linnane, *Biochem. Biophys. Res. Comm.*, 52, 398 (1968).
27. H. Küntzel and H. Noll, *Nature*, 215, 1340 (1967).
28. B. Kadenbach, *Europ. J. Biochem.*, 10, 312 (1969).
29. R. S. Criddle and G. Schatz, *Biochemistry*, 8, 322 (1969).
30. B. Ephrussi and P. P. Slonimski, *Comp. rend.*, 230, 685 (1950).
31. J. C. Mounolou, G. Perrodin, and P. P. Slonimski in *Biochemical Aspects of the Biogenesis of Mitochondria*, Adriatica Editrica, 1968, p. 133.
32. M. Rabinowitz, G. S. Getz, J. Casey, and H. Swift, *J. Mol. Biol.*, 41, 381 (1969).
33. W. D. P. Stewart, A. Haystead, and H. W. Pearson, *Nature*, 224, 226 (1969).
34. B. Weinstein, *Biochem. Biophys. Res. Comm.*, 35, 109 (1969).
35. R. P. Levine, *Science*, 162, 768 (1968).
36. J. K. Hoober, P. Siekevitz, and G. E. Palade, *J. Biol. Chem.*, 244, 262 (1969).
37. R. M. Smillie, D. Graham, M. R. Dwyer, A. Grieve, and N. F. Tobin, *Biochem. Biophys. Res. Comm.*, 28, 604 (1967).
38. B. G. Forget and S. M. Weissman, *J. Biol. Chem.*, 244, 3148 (1969).
39. R. H. Whittaker, *Science*, 163, 150 (1969).
40. L. Sagan (Margulis), *J. Theor. Biol.*, 14, 225 (1967).
41. L. Margulis, *J. Geol.*, 77, 606 (1969).

Origin of amino acid sensitivity in bacteria with RC^{rel} genotype

ᦒᦒᦒᦒ

L. ALFÖLDI

In *Escherichia coli* ribosomal and transfer RNA synthesis is under strict amino acid control. Auxotrophic bacteria deprived of the required amino acid stop immediately the synthesis of proteins and RNA. Borek, Rockenbach, and Ryan (1956) were the first to describe an auxotrophic strain of *E. coli* which continued to synthesize RNA for a considerable period even when starved for methionine. Stent and Brenner (1961) have demonstrated that a gene (RNA control or *RC* gene) is involved in the regulation by amino acids of RNA synthesis and that this gene can exist in two allelic states, stringent (RC^{str}) and relaxed (RC^{rel}). The location of this gene on the chromosome of *E. coli* was mapped between the streptomycin and histidine (Alföldi, Stent, and Clowes, 1962).

Many physiological consequences of the relaxed state are known (Edlin and Broda, 1968). One of the most typical amongst them is that on shift-down transfer—transfer of bacteria from a medium containing a complete complement of amino acids to a minimal medium containing only nutrients required for growth—bacteria with the RC^{rel} genotype start to grow considerably later than those with the RC^{str} genotype (Neidhardt, 1963; Alföldi et al., 1963). The longer delay in growth ensuing on down-shift transfer of bacteria of RC^{rel} genotype is a direct reflection of their relaxed control of RNA synthesis: whereas in stringent bacteria the transient amino acid starvation created by the shift-down stops RNA synthesis, in relaxed mutants RNA synthesis continues for a considerable time beyond the cessation of protein synthesis. This unbalanced biosynthesis of RNA by relaxed bacteria locks the cells into an aberrant state from which they return to the normal, balanced state very slowly.

Another physiological anomaly of the RC^{rel} genotypes is a marked amino acid sensitivity appearing at shift-down transfer (Alföldi et al., 1963). The presence of some supplementary, nonrequisite amino acids in the minimal medium at concentrations as low as 25–50 μg/ml inhibits the growth of relaxed bacteria at such a transfer. The natural amino acids to which bacteria of RC^{rel} genotype are so sensitive at shift-down transfer are L-serine, L-leucine, L-valine, L-isoleucine, L-cysteine, L-phenylalanine, and L-methionine. In a different strain a different spectrum of sensitivity may be present. On transfer from minimal to supplemented minimal medium, the same amino acids in the same concentration (except L-valine) are innocuous for the same bacteria and do not inhibit their growth. The sensitivity to valine of the strain *E. coli* K12 and its derivatives has long been known (Bonner, 1946), but valine-resistant RC^{rel} mutants may maintain shift-down valine sensitivity as before, demonstrating that the two kinds of valine sensitivity are of different origin (Alföldi, 1966). In this essay I should like to speculate on the possible origin of the shift-down amino acid sensitivity of the RC^{rel} bacteria.

Since all strains of *E. coli* with an RC^{rel} genotype hitherto examined have exhibited, without exception, shift-down amino acid sensitivity, it seemed simplest to postulate at first that the mutation of the RC^{str} gene to its RC^{rel} allele is pleiotropic. In fact, transfer of an RC^{rel} gene by conjugation results in the appearance of amino acid sensitivity in the recipient bacteria (Alföldi et al., 1963). But the same conjugation experiments have revealed that the spectrum of amino acid sensitivity of the recipient bacteria is often different from that exhibited by the RC^{rel} donor. This observation suggests that the phenotypic expression of sensitivity may be influenced by the presence of some other factors beside the RC^{rel} gene.

However, the appearance of an altered spectrum of amino acid sensitivity in the recipient bacterium may also be interpreted in another way. For example, it can be imagined that amino acid sensitivity is not so much a direct consequence of the RC^{rel} mutation, but rather that the

RC^{rel} genotype releases some inherent, but latent, regulatory anomalies of bacteria as a secondary effect of the RC^{rel} mutation.

This interpretation seems to be even more reasonable if it is recalled that relaxed bacteria undergo regulatory disturbances even at a simple down-shift transfer, since the coordination and economy of derepression become severely damaged because of the unbalanced RNA synthesis. Although this delicate state of unbalance is slowly overcome by the relaxed bacteria, it obviously may also create increased sensitivity to the influences which might be compensated by RC^{str} genotypes. Thus the shift-down amino acid sensitivity of the relaxed bacteria may be simply a manifestation of a deficiency in regulation, sensitive to unbalanced amino acid concentrations.

At present, neither the pleiotropic nature of the RC^{rel} mutation nor the validity of the latent sensitivity hypothesis has been proved. Nevertheless, there are some data which might help us to select between these two possibilities.

We might start our reasoning from the observation that the inhibitory action of any amino acid at shift-down transfer can be counteracted by different combinations of isoleucine, valine, leucine, and threonine; see Table 1 (Alföldi and Kerekes, 1964). This observation suggests that the origin of inhibition may be related to the biosynthesis of these amino acids. Or, more specifically, the origin of the amino acid sensitivity at shift-down transfer may reside in the derepression process, in the feedback inhibition susceptibility of individual enzymes of the pathway, or in both.

TABLE 1. SHIFT-DOWN TYPE OF AMINO ACID SENSITIVITY OF THE STRAIN *E. coli Hfr-C* RC^{rel} AND ITS NEUTRALIZATION BY OTHER AMINO ACIDS

Inhibitory amino acids	Counteracting combinations of amino acids
L-isoleucine	Valine
L-valine	Isoleucine
L-leucine	Isoleucine + valine
L-methionine	Isoleucine + valine
L-phenylalanine	Isoleucine + valine
L-cysteine	Isoleucine + valine + leucine
L-serine	Isoleucine + valine + leucine + threonine

But wherever the sensitive point(s) may be located in the biosynthetic pathway leading from aspartate through threonine to isoleucine, and jointly from pyruvate to valine and leucine (see Fig. 1), it is no exaggeration to say that, from the regulatory point of view, these are the pathways controlled by the most complicated interactions (Umbarger, 1961; Datta, 1969; Truffa-Bachi and Cohen, 1968). Therefore it is reasonable to expect that these pathways may contain sensitive points of regulation, which are only temporarily disturbed, and easily over-

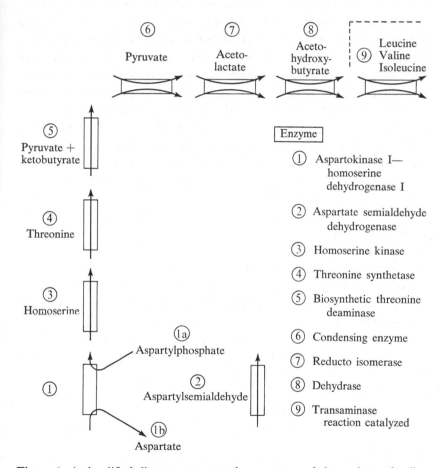

Figure 1. A simplified diagram representing enzymes of the pathway leading from aspartate through threonine to isoleucine and the interrelated pathways of valine and leucine.

come, when stringent bacteria are grown in the presence of slightly un-balanced concentrations of the end-product amino acids or of their structural analogues, but which are potentiated in the presence of the RC^{rel} gene.

We may expect such hidden regulatory anomalies of these pathways all the more since latent regulatory anomalies have been well known. For example, the strain *E. coli* K12 and its derivatives fail to grow in the presence of as much as a few micrograms per milliliter of valine, unless isoleucine is present in the medium (Bonner, 1946). Leavitt and Umbarger (1960) have demonstrated that this phenomenon is the result of increased feedback inhibition sensitivity of the condensing enzyme to valine. Thus even the slightest excess of valine brings about the inhibition of isoleucine synthesis and consequently leads to isoleucine starvation of the bacteria (Umbarger, 1961). The growth of *R. rubrum* is completely inhibited by a few micrograms per milliliter of threonine, and this effect is counteracted by isoleucine (Sturani et al., 1963). The origin of this anomaly in this bacterium lies probably in the increased feedback inhibition sensitivity of aspartokinase I and homoserine dehydrogenase I to L-threonine. Bacteria in the presence of a slight excess of threonine will starve for methionine and lysine. Isoleucine relieves this inhibition by a compensatory feedback control (Datta, 1969). (For additional examples see Gladstone, 1939; Rowley, 1953; Kerwar, Choldelin, and Parks, 1964.)

In any case, if we would like to pin-point the exact location of the shift-down amino acid sensitivity we must keep in mind that in complex medium the production of several key biosynthetic enzymes is repressed, that is, either they are present in the bacterial cell at a very much reduced level, or they are completely absent in such a milieu. Unfortunately, however, practically nothing is known about the exact events of the derepression (repressor-corepressor interactions) of the enzymes involved in the threonine-isoleucine-valine biosynthetic chain. Consequently, with many nonspecific effects operating, it is extremely difficult to generalize on the reduced enzyme levels of the repressed state in bacteria.

Nevertheless, there may be an indirect approach to the problem if shift-down amino acid sensitivity stems from the feedback inhibition sensitivities of some individual enzymes present at a reduced level in bacteria grown in complex media, rather than from repressor-corepressor interactions. In this case it would be expected that, since in a derepressed state enzyme levels are higher, amino acids inhibitory at shift-down transfer would be, in much higher concentrations, inhibitory also at a minimal-minimal transfer. Our experiments presently in progress have actually indicated that amino acids inhibitory in 25–50 μg/ml concentrations at shift-down also, in higher concentrations (500–1000 μg/ml or more), inhibit RC^{rel} bacteria at a minimal-minimal medium transfer. Inhibitions of this type can be counteracted with appropriate combinations of isoleucine, valine, leucine, and threonine, as at a shift-down transfer. In this way the shift-down amino acid sensitivity of the RC^{rel} bacteria can be reproduced at a transfer when derepression clearly does not play any role.

Gadó and Horváth (1965) have demonstrated that one *E. coli* strain with an RC^{str} genotype exhibits a down-shift type of amino acid sensitivity to low concentrations of leucine, isoleucine, serine, methionine, and threonine. However, the bacteria were inhibited by those amino acids only temporarily at such a transfer.

In our recent experiments we have shown that bacteria with an RC^{str} genotype are inhibited temporarily at a minimal-minimal transfer by the same high concentrations of the inhibitory amino acids (500–1000 μg/ml or more) which inhibit RC^{rel} genotypes. This temporary inhibition disappears in the presence of the same amino acid combinations which counteract the inhibitors in relaxed bacteria.

The observations that in stringent bacteria the same amino acid sensitivities are present as in bacteria with an RC^{rel} genotype, and that stringent bacteria counteract this inhibition spontaneously, suggests that in *E. coli* the pathway leading from aspartate through threonine to isoleucine and jointly to valine and leucine is particularly sensitive to unbalanced concentrations of the end-product and some related amino acids.

It seems reasonable to conclude, therefore, that a mutation of the RC^{str} gene to its RC^{rel} allele is not pleiotropic in itself but rather brings to light hidden regulatory shortcomings of some biosynthetic pathways. Should this interpretation be correct, then bacteria with an RC^{rel} genotype would be particularly useful objects, not only in the study of regulatory problems of RNA biosynthesis, but also in the investigation of regulatory problems of biosyntheses in general.

REFERENCES

Alföldi, L. (1966). *Genetika (Moscow)*, 1, 105.
Alföldi, L., and E. Kerekes (1964). *Biochim. Biophys. Acta*, 91, 155.
Alföldi, L., G. S. Stent, and R. C. Clowes (1962). *J. Mol. Biol.*, 5, 348.
Alföldi, L., G. S. Stent, M. Hoogs, and R. Hill (1963). *Z. Vererbungslehre*, 94, 285.
Bonner, D. M. (1946). *J. Biol. Chem.*, 166, 545.
Borek, E., J. Rockenbach, and A. Ryan (1956). *J. Bact.*, 71, 318.
Datta, P. (1969). *Science*, 165, 556.
Edlin, G., and P. Broda (1968). *Bact. Rev.*, 32, 206.
Gadó, I., and I. Horváth (1965). *Acta Microbiol. Acad. Sci. Hung.*, 12, 59.
Gladstone, G. P. (1939). *Brit. J. Exp. Path.*, 20, 189.
Kerwar, S. S., V. H. Choldelin, and L. W. Parks (1964). *J. Bact.*, 88, 179.
Leavitt, R., and H. E. Umbarger (1960). *Bact. Proc.*, p. 183.
Neidhardt, F. C. (1963). *Biochim. Biophys. Acta*, 68, 365.
Rowley, D. (1953). *J. Gen. Microbiol.*, 9, 37.
Stent, G. S., and S. Brenner (1961). *Proc. Nat. Acad. Sci. U.S.*, 47, 2005.
Sturani, E., P. Datta, M. Hughes, and H. Gest (1963). *Science*, 141, 1053.
Truffa-Bachi, P., and G. Cohen (1968). *Ann. Rev. Biochem.*, 37, 79.
Umbarger, H. E. (1961). *Cold Spring Harbor Symp. Quant. Biol.*, 26, 301.

The transfer of biological stress in microorganisms

ERNEST BOREK

If you are lucky enough to have lived in Paris as a young man, then wherever you go for the rest of your life it stays with you.

ERNEST HEMINGWAY

It was boredom with biochemistry that brought me to Lwoff. In the late 1940's the main preoccupation of that science was with tracing, atom by atom, the origins of various intermediary metabolites. I continued some research perfunctorily, but my main interest was writing a book envisioned as "Biochemistry for the Millions." (It turned out to be, in those pre-Sputnik days, for the dozens.) I wrote in that book, "So far the biochemist's efforts have been directed mainly toward the completion of the atlas of molecular anatomy," and went on to say, "We must begin to assemble patiently still another science, molecular physiology—the study of the relation between molecular structure and cellular function" (Borek, 1952a). I was ready to become a molecular biologist before the term was coined.

Writing the book had another advantage besides allowing me now to prove that my foresight is not hindsight: I read a great deal. I was struck by a sentence of Sir Frederick Gowland Hopkins, the great Cambridge biochemist who was essentially the founder of modern biochemistry. Hopkins said, "In exploring and cultivating the fields of nature, the chemists were best provided with the machinery for this cultivation, but the biologist knew best the lay of the land."

It was the late Heinrich Waelsch who showed me a two-sentence abstract of a paper on lysogenic induction that André Lwoff presented at some meeting in America sometime in 1950. A few days in the library spent reading Lwoff's previous work convinced me that I had my biologist and my phenomenon. For induction provided the cardinal requisites for a biochemical exploration of a biological phenomenon: it was reproducible and nearly homogenous, and masses of material were available. Accordingly, I applied for support to work in Lwoff's laboratory. I thought it was presumptuous to announce my plans before I had the financial means to do so, and therefore I delayed until April my letter announcing my desire to come sometime that fall.

As I heard it told later, my plans were less than enthusiastically received. Fortunately, Melvin Cohn overheard the tirade about the *cochon* of an American who thought he could come on a few months' notice, and he, being a former student of mine, interceded on my behalf —an example of a student developing the career of his professor.

Great contributions to science come from two sources. There are those who, by keen observation, see something never seen before by the eyes of man. And then there are those, the more gifted ones, I think, who take something seen by all but see it differently and give it new meaning. Lwoff, of course, belongs to the latter few. But extraordinary originality in science exacts a price. Most contemporaries refuse or are unable to see what is revealed by the insight of genius. A world-renowned biologist told me I was a fool for trying to study lysogenic induction: it was a trivial, evanescent phenomenon which would contribute no more to biology than did the "fertilization" of sea urchin eggs induced by stroking them with a camel's-hair brush. A scientist high in the French administrative echelons asked me how I enjoyed working with Lwoff. When I replied with enthusiasm and proffered the opinion that Lwoff is one of the great biologists of the century, the gentleman snorted, "Not the least of his abilities is the one to impress Americans."

Nor did Lwoff's quarters fail to testify to the lack of contemporary appreciation of his work. I was appalled by the crowded, dilapidated attic into which the Microbial Physiology Department was packed. (The best and most spacious laboratories were assigned to a biochemist the importance of whose contributions, the effect of enormous pressures on enzymes, was immediately apparent to everyone, especially to deep-sea fish.)

But what a joyous place was that attic! Ideas on science, world affairs, literature, and aesthetics lapped one like a beneficent, warm sea. I was fortunate in sharing a tiny laboratory with the late Francis Ryan, who, with his wife Betty, patiently taught me the simplest rudiments of microbiology. Therefore, I was quickly able to adhere to Lwoff's dictum: "Work only with bacteria in logarithmic growth phase." (I was appalled to realize that I had published extensively on metabolic inhibitors of bacteria with inocula which were probably 99% dead.)

A study of inaptitude—the reduced capacity for induction after deprivation of glucose—seemed to be a pliant area for an entering wedge by some chemical tools. My assessment of the source of inaptitude as being due to the accumulation of inhibitors of phage formation was not applauded by Lwoff: "You Americans have inhibitors on the brain," he said.

I did not dare entertain the inhibitors until I returned home. My first approach to the inhibitors accumulating in starved lysogenic organisms was prompted by my readings for my book. I was impressed by Otto Loewi's demonstration of the existence of a biological entity by stimulating one frog heart and transferring the stimulus to another heart via the juices of the first one. I thought one could design similar experiments with donor and recipient bacteria. We had established in Lwoff's laboratory—he graciously declined coauthorship—that amino acid deprivation also produces inaptitude (Borek, 1952b). Therefore, I starved a lysogenic, amino acid auxotroph, eliminated the organisms by filtration, and suspended in the filtrate wild-type lysogenic *Escherichia coli* K12. And, lo and behold, a large fraction of the organisms, for which that medium was complete, survived the usual inducing dose! Triumph supreme! Despite the deprecation of my mentor in biology I had my inhibitors.

My naive jubilation did not last long. Through a torturous path, details of which I will spare the reader, I came to the conclusion that starvation media probably did not contain inhibitors but certainly they contained ultraviolet-absorbing materials which screened out the inducing UV energy. That I had not realized this earlier probably sounds amateurish now, but it must be realized that in 1952 very few valid quantitative experiments had been performed on the effects of starvation on bacteria. Indeed, a search of the literature revealed only a very brief, posthumous note by Marjorie Stephenson on the excretion of nucleosides by bacteria during starvation of phosphate (1949).

The excretion of nucleic acid fragments during starvation naturally led to a study of the *internal* nucleic acid content of the bacterial cell, for by then I was deeply immersed in a hypothesis of internal screening as the source of inaptitude. And, of course, that is how we discovered what is now called the "relaxed and stringent" control over the synthesis of RNA, a phenomenon so extensively studied and reviewed that I will not waste the reader's time with it except to point out that, like all novel findings, it had to struggle for its existence. The *Archives of Biochemistry and Biophysics* rejected the manuscript as concerning a trivial phenomenon which is merely a single exception to the general rule, and, adding insult to ignorance, the editor went on to say that my paper was poorly written! And this was at a time when I worked in a small laboratory with two assistants, so that I had time to polish the sentences of our rare communications until they shone (Borek, Ryan, and Rockenbach, 1955).

I took comfort from the history of science: because of a difference of opinion with a dean I was exiled to an obscure laboratory in the Veterans Administration Hospital of Brooklyn, and my work was not appreciated. I was bound to be on the right track.

I made one more attempt at what might be called an Otto Loewian experiment. Since bacteria lose so many metabolites during starvation, is it not possible that inaptitude is due to the loss of some UV-sensitive target? I therefore tried to incubate noninduced lysogenic organisms (and their protoplasts) in the lysates of induced ones. Unfortunately the results were not consistently negative. We filled notebooks with marginally meaningful experiments. To our credit I must add that they were never published. I would cheer the lagging spirits of my two predoctoral associates with the assessment that in our present state of ignorance about molecular mechanisms in a cell we do not need an IBM machine to find a phenomenon.

And we found one. Ideas for Otto Loewian experiments come under unusual circumstances. Loewi dreamt his; mine occurred to me while I was skiing. (What one sees among the youngsters in American ski lodges may have been the stimulus.) I decided to try to pass the effects of UV irradiation to an uninduced $F^- \lambda+$ organism via conjugation with an irradiated $F^+ \lambda-$. The scheme worked, and a new type of induction, "indirect induction," was added to the tools for the investigation of lysogeny. (Drs. Calef and Devoret have graciously called the phenomenon the Borek-Ryan effect. If the reader is interested, I can refer him to a review by the latter: Devoret and George, 1967.)

The mechanism of indirect induction is by no means clarified. (Since X-ray induction is not transferred but $F+$ organisms exposed to *both* X and UV irradiation are good donors for indirect induction, I cannot accept the conclusion that it is merely a damaged DNA which is the active agent in indirect induction.) However, my attention and limited facilities became focused on another area: the methylation of nucleic acids. The discovery of this mechanism came about through an extraordinary coincidence. We observed the relaxed control over RNA synthesis, that is, the continuing synthesis of RNA in the absence of the essential amino acid, in a methionine-requiring auxotroph *E. coli* K12 W6. (The archetypal organism isolated by Tatum was designated 58-161 by him. The mutation to relaxed control occurred later during Lederberg's manipulation of it.) When Dunn (1959) showed the presence of a variety of methylated bases in tRNA, I was certain that there was something wrong with the RNA which accumulated during methionine starvation. The methyl groups substituted on the nitrogens must come from methionine. My graduate student at Columbia, Mr. Lewis R. Mandel, showed that indeed the RNA which accumulates is methyl deficient and that all the methyl groups, even that for ribosylthymine, come from methionine. Then, with methyl-deficient RNA as a tool, another student, Mr. Erwin Fleissner, showed the existence of enzymes which complete the synthesis of the macromolecular structure of tRNA by adding methyl groups with fantastic precision.

Life became hectic after these discoveries. A number of bright young men entered the field and went through it like locusts, leaving behind mounds of publications. The net purport of these was that methylation of nucleic acids serves no function whatever. The source of the confusion was rather obvious. The organism in which we found the uncoupling of RNA synthesis and methylation must be grown to a high population in the presence of methionine before it can be starved of the amino acid, and therefore the subsequent extraction of tRNA yields a mixture of methylated and methyl-deficient product. With such mixtures and crude enzyme systems nothing of value could come from quick *in vitro* studies. I spent a great deal of energy in those days convincing my young associates that we must concentrate on meaningful biological experiments and not waste time on meaningless but publication-yielding "quickies." (Gresham's law—bad money forces the good out of circulation—applies, alas, in science as well.)

From the work of several laboratories with clean systems we know today that methylation of tRNA affects its interaction with the amino-acid-charging enzymes (Shugart et al., 1968), its coding response (Peterkofsky, Jesensky, and Capra, 1966), and its ability to attach to ribosomes (Gefter and Russell, 1969). That the methylation of DNA serves at least one cognitive role has also been shown unequivocally (Arber and Linn, 1969).

I do not wish to dwell here on the methylation of nucleic acids. All too many reviews are available. I will discuss only one aspect of their possible biological involvement because Lwoff had a remote influence on its development.

In early 1951 Dr. Lwoff would drop into my laboratory frequently and would pick up a tube of *Bacillus megatherium* or of *E. coli* K12. He would look at the contents intently, purse his lips, and say, "These bacteria have cancer." Then he would walk away.

The analogy had not occurred to me until then, but thereafter I would glance sporadically at the mire called "cancer literature." Therefore, I knew enough of the background of alkylating carcinogens to be arrested by the report of Magee and Farber (1962) on the alkylation of nucleic acids by the chemical carcinogen dimethylnitrosamine. From the work of Sir Alexander Hadow, we knew that alkylating carcinogens alkylate DNA aberrantly. But here was evidence that tRNA was alkylated perhaps ten times as much. We had assumed causality between carcinogenesis and alkylation of DNA simply because the alkylation of tRNA and indeed tRNA itself were unknown at the time of the original discovery.

I knew by then that all nucleic-acid-methylating enzymes are species specific. Therefore, I proposed a hypothesis of carcinogenesis by methylating enzymes foreign to the host, the information for whose synthesis could be introduced by an oncogenic virus. My preference was for the tRNA methylases: there are more of them, and I already knew of their modulation in phage infection and induction (Wainfan, Srinivasan, and Borek, 1965).

The hypothesis has been tested in several laboratories, and the crude tRNA methylases have been found to be aberrantly altered so that they can recognize more sites for methylation than their normal counterparts in thirty different tumors of both human and animal origin (Borek, 1969). Of these thirty the situation with the tumor induced by the alkylating carcinogen ethionine is the most curious. It has now been shown unequivocally by Ortwerth and Novelli (1969) that this agent ethylates only tRNA, *not* DNA, of the target organ, the liver. However, the tRNA methylases of the liver become aberrant before any cytological changes are visible (Hancock, 1968).

Another test of the hypothesis is an examination of the population of tRNA's in tumor cells: the methylation of the tRNA's themselves are now known to be aberrantly excessive in twelve different tumors. Moreover, novel tRNA's, in respect to their physicochemical properties as shown by elution profiles, have been observed in several others (Baliga et al., 1969).

The meaning of all this in terms of regulation is not yet clear. Of course, tRNA's can be determinants in protein synthesis in both bacterial and mammalian systems (Anderson and Gilbert, 1969). We are now certain that a component of the protein-synthesizing machinery of the tumor cell, tRNA, is *qualitatively* different from the population of normal tRNA's.

One could speculate on the presence in cancer cells of forbidden tRNA's, so that certain proteins which must be in limited abundance in normal cells are continuously synthesized. Or one could visualize the elimination of certain functional tRNA's by alteration so that certain proteins whose presence is *essential* become in short supply. A third possibility is that the enzymes which we assay collectively as the tRNA methylases may also methylate aberrantly some cytoplasmic nucleoprotein which, on return to the nucleus, serves as a signal for dedifferentiation.

164 ERNEST BOREK

But the harvest of glib hypotheses is always abundant. Crucial probes for their validation, on the other hand, are rare flowers; their cultivation requires time, energy, and ingenuity.

May my mentor in biology yet see whether this trail started in his laboratory two decades ago, with the bacteria which "have cancer," leads anywhere.

REFERENCES

Anderson, W. F., and J. M. Gilbert (1969). *Biochem. Biophys. Res. Comm.,* 36, 456.
Arber, W., and S. Linn (1969). *Ann. Rev. Biochem.,* 38, 467.
Baliga, B. S., E. Borek, P. R. Srinivasan, and I. B. Weinstein (1969). *Proc. Nat. Acad. Sci. U.S.,* 62, 899.
Borek, E. (1952a). In *Man the Chemical Machine,* New York, Columbia University Press.
Borek, E. (1952b). *Biochim. Biophys. Acta,* 8, 211.
Borek, E. (1969), p. 163. In *Exploitable Molecular Mechanisms and Neoplasia,* Williams & Wilkins Co., Baltimore.
Borek, E., A. Ryan, and J. Rockenbach (1955). *Bact.,* 69, 460.
Devoret, R., and J. George (1967). *Mutation Res.,* 4, 713.
Dunn, D. B. (1959). *Biochim. Biophys. Acta,* 34, 286.
Gefter, M. L., and R. L. Russell (1969). *J. Mol. Biol.,* 39, 145.
Hancock, R. L. (1968). *Biochem. Biophys. Res. Comm.,* 31, 77.
Magee, P. N., and E. Farber (1962). *Biochem. J.,* 83, 114.
Ortwerth, B. J., and G. D. Novelli (1969). *Cancer Res.,* 29, 380.
Peterkofsky, A., C. Jesensky, and J. D. Capra (1966). *Cold Spring Harbor Symp. Quant. Biol.,* 31, 515.
Shugart, L., B. H. Chastine, G. D. Novelli, and M. P. Stulberg (1968). *Biochem. Biophys. Res. Comm.,* 31, 404.
Stephenson, M., and J. M. Moyle (1949). *Biochem. J.,* 45, VII.
Wainfan, E., P. R. Srinivasan, and E. Borek (1965). *Biochemistry,* 4, 2845.

Of man and pneumococcus, an historical paradox

ROBERT AUSTRIAN

In the fifty years spanning the scientific career of André Lwoff, few organisms have played a larger role than has pneumococcus in contributing to the understanding of bacterial disease and of the biology of the cell. Historically it serves as a link between France and the United States, having been described for the first time in the year 1881 by Pasteur in Paris and by Sternberg in New York. It was not until several years later, however, that its importance as the major cause of bacterial pneumonia in man was established by Fraenkel (1884) and by Weichselbaum (1886). Writing of pneumonia at the turn of the century, Osler (1909), in his epochal textbook entitled *The Principles and Practice of Medicine,* designated it "Captain of the men of death."

Study of the causative organism of a disease of such importance, which strikes down the young as well as the aged, was bound to be intense. How else could a rational approach to therapy be achieved? The initial attempts to produce an effective antiserum in a fashion analogous to that employed so successfully for diphtheria antitoxin failed to meet with success, for the serologic diversity of pneumococcus had not been recognized. Only after three distinct serotypes and an heterogenous collection of strains categorized as Group IV (Dochez and Gillespie, 1913) had been identified was an approach to the serotherapy of pneumonia possible. The beneficial effect of Type I antipneumococcal horse serum was established between 1910 and 1920.

As a result of the foregoing investigations, it had become apparent that the capsular antigens of the pneumococcus played a major role in the virulence of the organism and were responsible for its antigenic diversity. Biochemical study of these capsular substances led to the discovery that they are complex polysaccharides (Heidelberger and Avery, 1923) and to the milestone in immunology that molecules other than proteins can stimulate the formation of antibodies. Attempts to define better the conditions determining reacquisition by avirulent pneumococci of the ability to produce a capsule resulted in the first demonstration of bacterial transformation by Griffith (1928). Although the significance of Griffith's findings was not appreciated fully at the time, they were to prove to be the initial step in the establishment of deoxyribonucleic acid as the major biochemical determinant of heredity (Avery, MacLeod, and McCarty, 1944). These observations serve well to illustrate the "fallout" of fundamental biological knowledge resulting from the study of an important pathogen of man.

Meanwhile, further progress was being made in the classification of pneumococcal capsular types and in the production of sera to treat pneumococcal disease. Definition of 32 serotypes by Cooper and her associates (1929) and the substitution of rabbit antisera for those produced in horses marked the next steps in the efforts to provide effective therapeutic measures for the management of pneumonia. Despite these advances, the era of serotherapy was nearing its end. In 1935, Domagk reported the initial use of a sulfonamide in the treatment of a bacterial infection; and, in 1938, Whitby described the successful employment of sulfapyridine for the therapy of pneumococcal disease. For the first time, treatment of pneumococcal infection, regardless of the capsular type of the causative organism, was possible with a single chemotherapeutic agent.

Although the death rate from pneumonia in the United States had begun to fall at the beginning of the twentieth century and had continued to do so except in the years of the influenza pandemic of 1918–1919, the total number of deaths from pneumonia continued at a significant level as a result of the steady growth of the nation's population. No more than a slight effect of serum therapy on the death rate from pneumococcal disease can be observed, a not too surprising finding in view of the fact that such treatment was generally available for the management of infections caused by only a few capsular types. Since sulfonamides are potent against pneumococcus regardless of type, their impact on the death rate from pneumonia, following their widespread introduction in 1938, is clearly evident. Fear of the potentially fatal outcome of pneumococcal infection and the lower but continuing significant death rate accompanying sulfonamide therapy led for a time to retention of the highly efficient laboratory techniques for the isolation and typing of pneumococci and, on occasion, to the treatment of the seriously ill with both sulfonamides and type-specific antipneumococcal serum. The advent of penicillin, however, was soon to bring about marked changes in the diagnostic procedures employed in the management of pneumonia and in medical attitudes regarding its prognosis.

The successful treatment of pneumococcal pneumonia with penicillin was reported by Tillet and his associates in 1945. Employing doses that are miniscule by today's standards, they achieved highly satisfactory results. Resembling the sulfonamides in its potency against pneumococci of all capsular types, penicillin differed from the former drugs in being bactericidal rather than bacteriostatic and in manifesting significantly lesser toxicity. To many physicians, it appeared that the battle to control and, perhaps, even to eliminate pneumococcal disease had been won.

The clear superiority of penicillin over all forms of treatment for pneumococcal infection available previously led quickly to cessation of the production of therapeutic type-specific antipneumococcal sera. Diagnostic pneumococcal typing sera, by-products of therapeutic sera, became decreasingly available; and, by the early 1950's, no commercial source of such diagnostic reagents could be found in the United States. For more than a decade, their only commercial source in the Western world has been the Statens Seruminstitut, Copenhagen, Denmark. For the identification of pneumococcus in clinical laboratories, reliance was placed on bacterial cellular and colonial morphology, the solubility of the organism in bile salts, and its sensitivity to ethylhydrocupreine hydrochloride. None of these tests permitted the identification of pneumococcus with the speed or accuracy of the capsular precipitin or *Quellung* reaction. Neither did they permit the useful prognostic inferences that could be drawn from knowledge of the capsular type of pneumococcus responsible for infection.

With the abandonment of time-tested methods, the frequency and accuracy with which pneumococcus was recognized in the clinical laboratory declined rapidly. These changes are reflected in the medical writing of the period. In 1950, Reimann wrote: "Unfortunately, since the determination of types of pneumococci which cause pneumonia is no longer of practical necessity for therapy, the procedure has been almost entirely abandoned." Six years later, this view was echoed by Rhoads (1956): "A revealing sign of the times is that no efforts were made at typing pneumococci in the entire series, for specific antiserum has not been used at this hospital for many years." An even more nihilistic outlook is found in the statement of Bridge in 1956: "In cases of pneumonia in the age group 15–60 without significant coexistent disease, it is felt that routine bacteriologic studies are unnecessary." The situation with regard to pneumococcal typing in Great Britain would appear to be similar to that in the United States, for an editorial on pneumococcal pneumonia published in *Lancet* in 1968 states: ". . . so far as we know, no one types them [pneumococci] routinely."

At the same time that penicillin was undergoing its initial trials in the treatment of pneumococcal disease, prevention of such infection was under scrutiny elsewhere. In an effort to reduce the loss of manpower caused by epidemic pneumococcal pneumonia in an army air force training school, MacLeod and his associates (1945) immunized 8500 men against four of the six pneumococcal types responsible for the epidemic disease. A similar group of men served as controls. The vaccine consisted of 50 μg each of the purified capsular polysaccharides of pneumococcal types I, II, V, and VII. Immunization was not provided against the other two epidemic types, IV and XII, which served as controls in both groups. The results of the vaccine trial showed clearly that infection caused by the pneumococcal types included in the vaccine could be prevented by immunization with them. The study suggested also that, in a closed population such as the one under investigation, the presence of immunity in half the group impeded the spread of epidemic strains through the population with a resultant lowering of the incidence of the disease in the controls. The effects of the vaccine were quite specific, for there was no reduction in the incidence of disease caused by pneumococcal types other than those included in the vaccine in either segment of the population.

That pneumococcal vaccine could also prevent infection in an older population was reported several years later by Kaufman (1947), who demonstrated a 90% reduction in the incidence of pneumonia and of bacteremia caused by pneumococcal types I, II, and III in a controlled study involving 10,000 persons 50 years of age or older, half of whom received vaccine. Further studies of vaccines of purified pneumococcal capsular polysaccharides showed that, when 50 μg each of six different polysaccharides were administered simultaneously, most subjects manifested an antibody response to all six (Heidelberger, MacLeod, and DiLapi, 1948); that half-maximal levels of antibody persisted as long as 5–8 years after a single injection of capsular polysaccharides (Heidelberger, 1953); and that injection of the vaccine was followed by a rate of untoward reactions considerably lower than that following the administration of typhoid vaccine. For that definable segment of the population at high risk of infection and of death caused by the commoner and more invasive pneumococcal types, the polyvalent vaccine of purified capsular polysaccharides would appear to have possessed the attributes of a highly satisfactory prophylactic agent.

Although licensed for commercial production after World War II, pneumococcal vaccine became available at that historical point in time when the medical profession had become convinced that pneumococcal disease was no longer a significant problem, and use of the vaccine was negligible. It was withdrawn from the market in the early 1950's; and, at the same time, the manufacturer asked that the license for its production be revoked without prejudice.

The foregoing historical notes delineate an extraordinary paradox. Having obtained, after years of intensive search, the means to treat effectively or to prevent a common infection with a significant fatality rate, the medical profession abandoned forthwith the technique of diagnosing it accurately. Failure to recognize pneumococcus in the hospital laboratory led soon to the widely held view that pneumococcal disease had become a rarity and to the impression, for example, that only three or four cases of pneumococcal pneumonia were being admitted annually to the 3000 bed municipal hospital of a large urban area.

To determine whether or not prevailing concepts concerning the incidence and severity of pneumococcal disease and the distribution of infecting pneumococcal types were indeed true, a laboratory was established in 1952 to serve a large urban medical service to which previously untreated patients were admitted in large numbers, including those ill with pneumonia. Cultures of respiratory secretions and of blood were obtained routinely before the administration of antimicrobial agents and were examined by previously established techniques for the isolation of pneumococcus. All strains of bacteria isolated which resembled pneumococci were typed by the capsular precipitin or *Quellung* technique.

The results of this investigation of pneumococcal infection (Austrian and Gold, 1964) established a number of facts:

1. That 20–25 bacteremic pneumococcal infections were observed annually for each 100–125 medical ward beds, an incidence similar to that seen before the advent of antibiotics.

2. That nonbacteremic pneumococcal pneumonia continued to occur with a frequency 3–4 times that of bacteremic infection.

3. That, with the exception of capsular Type II, the types of pneumococcus responsible for bacteremic pneumococcal infection were the same as those causing such infections before antibiotics were available.

4. That the fatality rate resulting from bacteremic pneumococcal pneumonia treated with penicillin or with other antibiotics was 17–18%, and that, in persons 50 years of age or older or in those with underlying systemic illness, the fatality in similarly treated patients exceeded 25%.

5. That the preponderance of deaths occurred in those sustaining irreversible physiologic damage within 5 days of the onset of infection.

The findings indicated clearly that pneumococcal disease is still prevalent, and that, for certain definable segments of the population, such infection is still accompanied by a significant risk of a fatal outcome, a risk unlikely to be lessened by antibacterial therapy alone. Until the mechanisms whereby pneumococcus damages man and the means to control them are known, the discarded technique of prophy-laxis would appear to offer the greatest likelihood of reducing the still significant mortality from pneumococcal infection.

"Pneumonia and influenza" is the only category of infectious dis-ease among the ten leading causes of death in the United States, rank-ing fifth in 1965. In that year 59,608 deaths were attributed to pneu-monia alone (*Statistical Abstract of the United States,* 1967). If an overall mortality rate for pneumonia of 10% is assumed, there were over half a million such illnesses in the same year. In periods when influenza virus is not epidemic, the preponderance of deaths from pneumonia is due to bacterial infection, and at least 80% of these are caused by pneumococcus. In addition to the saving in lives that could be effected by prophylactic vaccination against infection with the more prevalent and virulent pneumococcal types, the reduction in human disability that might result from the prevention of such illness would be appreciable. Because the incidence of pneumonia is higher in the aged, the problems it causes can be expected to increase. In 1966, there were 35,718,000 persons 55 years of age or older in the United States. The number falling into this same age group is projected by the Bureau of the Census (*Statistical Abstract of the United States,* 1967) to increase to 46,214,000 in 1985. It would seem reasonable, in the light of these estimates, to attempt to reduce the cost in lives and in illness brought about by pneumonia by reintroduction of polyvalent pneumococcal vaccine.

To further this end, a program designed to lead to the relicensing of pneumococcal vaccine is currently in progress, supported by the National Institute of Allergy and Infectious Diseases of the U.S. Public Health Service. Surveillance of the pneumococcal types responsible for bacteremic infection has been extended geographically, and more than 1500 bacteremic infections have been identified in nine urban areas of the United States in the last 30 months (Austrian, 1969). Types I, III, IV, VII, VIII, and XII account for 52.5% of such infections.

Surveillance of the incidence of pneumococcal infection in popula-tions of defined size is now also in progress. The development of a radio-immunoprecipitin assay for antibody to pneumococcal capsular polysaccharides (Schiffman, 1969) makes possible for the first time a rapid and convenient method for measurement of the antibody response to pneumococcal infection in patients with and without bacteremia. Demonstration of such a response is an essential part of establishing the causal role in infection of an organism isolated from sputum.

Vaccine is currently in production. After having been tested for safety and for efficacy in inducing an immune response, it will be assayed for its ability to prevent disease in field trials in the populations in which the incidence of pneumococcal disease has been ascertained previously. If the results are similar to those obtained by MacLeod and his coworkers (1945), as anticipated, vaccine will be made generally available once again.

The medical profession has not been immune to fads in the past, nor is it likely to be so in the future. Significant therapeutic advances are not infrequently accompanied by overoptimism and the hope that a particular disorder can be eliminated completely. This latter goal is rarely, if ever, attained where illnesses of infectious origin are concerned, however, and control of such disorders is more likely to be achieved by immune prophylaxis than by the administration of antimicrobial drugs. Let us not lay ourselves open to the charge of Molière: ". . . et j'ai connu une homme qui prouvait, par de bonnes raisons, qu'il ne faut jamais dire: 'Une telle personne est morte d'une fièvre et d'une fluxion sur la poitrine'; mais: 'Elle est morte de quatres médecins et de deux apothocaires' " (L'Amour Médecin, Acte II, Scène I).

REFERENCES

Austrian, R. (1969). Unpublished observations.

Austrian, R., and J. Gold (1964). *Ann. Int. Med.,* **60,** 759.

Avery, O. T., C. M. MacLeod, and M. McCarty (1944). *J. Exp. Med.,* **79,** 137.

Bridge, R. G. (1956). *Dis. Chest,* **30,** 194.

Cooper, G. M., C. Rosenstein, A. Walter, and L. Peizer (1932). *J. Exp. Med.,* **55,** 531.

Dochez, A. R., and L. J. Gillespie (1913). *J.A.M.A.,* **61,** 727.

Domagk, G. (1935). *Deutsche med. Wochnschr.,* **61,** 250.

Editorial (1968). *Lancet,* **1,** 1360.

Fraenkel, A. (1884). *Verhandlungen des Kongress für innere Medicin. Dritter Kongress,* vol. 3, p. 17.

Griffith, F. (1928). *J. Hyg.,* **27,** 113.

Heidelberger, M. (1953). In *The Nature and Significance of the Antibody Response,* edited by A. M. Pappenheimer, Jr., New York, Columbia University Press, p. 90.

Heidelberger, M., and O. T. Avery (1923). *J. Exp. Med.,* 38, 73.

Heidelberger, M., C. M. MacLeod, and M. DiLapi (1948). *J. Exp. Med.,* 88, 369.

Kaufman, P. (1947). *Arch. Int. Med.,* 79, 518.

MacLeod, C. M., R. G. Hodges, M. Heidelberger, and W. G. Bernhard (1945). *J. Exp. Med.,* 82, 445.

Osler, W. (1909). *The Principles and Practice of Medicine,* 7th Ed., New York, D. Appleton & Co., p. 165.

Pasteur, L. (1881). *Bull. Acad. Méd. (Paris),* 10, 2e sér., 94.

Reimann, H. A. (1950). *Ann. Int. Med.,* 33, 1246.

Rhoads, P. S. (1956). *Med. Clin. N. Amer.,* 40, 89.

Schiffman, G. (1969). Unpublished observations.

Statistical Abstract of the United States, 1967, 88th Ed., Washington, D.C., U.S. Dept. of Commerce.

Sternberg, G. M. (1881). *Nat. Bd. Health Bull.,* 2, p. 781.

Tillet, W. S., J. E. McCormack, and M. J. Cambier (1945). *J. Clin. Invest.,* 24, 589.

Weichselbaum, A. (1886). *Medizinische Jahrbücher,* 1, 483.

Whitby, L. (1938). *Lancet,* 1, 1210.

The evolution of an infectious disease of man

~~~~~~

A. M. PAPPENHEIMER, JR.

En matière de l'évolution, où les documents sont forcément rares sur lesquels on peut s'appuyer, un fait bien établi, fût-il unique, possède une importance capitale, et, dans ce domaine, il convient de passer sans trop de scrupules du particulier au général. C'est le privilège, et aussi l'excuse, des conceptions théoriques de poser de multiples problèmes et de susciter de nouvelles recherches.

A. LWOFF, *L'évolution physiologique*, 1943

I first met André Lwoff at the memorable Cold Spring Harbor Symposium of 1946, the first to be held after World War II. After the meeting, he and Jacques Monod spent a weekend at our country place in Scotland, Connecticut. I have always had great affection and admiration for France and for the French, and I was due for a sabbatical leave from New York University a few years later. André generously agreed to save a place for me in his Service de Physiologie Microbienne à l'Institut Pasteur during the year 1951–1952. In retrospect, I think that year spent with my family in Paris was probably the most satisfying and happiest of my life.

A bacteriophage is a nucleoproteinic particle adorned with a tail.

A. LWOFF, *Journal of General Microbiology*, 1957

We arrived in Paris in June 1951, soon after Lwoff and Gutmann (1950) had demonstrated, by observations on single cells in microdrops, that bacteriophages found in cultures of lysogenic bacteria arise as the result of the spontaneous induction and lysis of occasional bacteria. Lwoff, Siminovitch, and Kjeldgaard (1950) had just defined "prophage" and had demonstrated its presence in all cells of a lysogenic culture by showing that after exposure to a mutagenic stimulus, such as ultraviolet irradiation, phage multiplication could be induced in every cell with subsequent lysis and release of mature phage. Not long after our arrival in Paris, the paper by Freeman (1951) appeared, describing the simultaneous conversion of certain nontoxigenic diphtheria bacilli to lysogeny and toxinogeny.

I knew very little about bacteriophages at that time and almost nothing about lysogeny. I did, however, have a consuming interest of long standing in the biology of diphtheria and in the organism that causes it. A letter from my student, Lane Barksdale, who wrote that he had decided to study, for his doctoral thesis, the relationship of phage to the pathogenicity of the diphtheria bacillus, together with the exciting observations of André Lwoff, was all the additional encouragement needed to arouse my curiosity in the mechanism underlying this unusual lysogenic conversion. In this paper I shall attempt to show how complex the series of evolutionary circumstances must have been that led, ultimately, to the human infectious disease known as diphtheria.

Examinés sous un jour particulier, des phénomènes autrefois sans signification vont prendre toute leur valeur et contribuer a la solidité de l'édifice.   A. LWOFF, *L'évolution physiologique*, 1943

Shortly after Loeffler had established the etiology of diphtheria and had shown experimentally that diphtheria bacilli isolated from the throats of patients were capable of causing a similar disease in guinea pigs, rabbits, and pigeons, he noted (1887) that organisms with the same distinctive morphology could be isolated from the throats of normal, healthy individuals. Not infrequently, these organisms were indistinguishable, in all respects tested save one, from those isolated from patients with the disease; they were "avirulent" for guinea pigs. It was, of course, established at about the same time by Roux and Yersin (1888) that the lesions in diphtheria are caused by a soluble, heat-labile extracellular toxin. Organisms that fail to produce specific toxin cannot cause true diphtheria.[1] Later, nontoxigenic corynebacteria, closely resembling *Corynebacterium diphtheriae,* were isolated from the nasopharynges of horses, cows, and sheep (Andrewes, 1923), and occasionally typical toxin-producing strains were found in domestic animals.

In 1921, Moss, Guthrie, and Gelien made a survey of a Baltimore orphanage that had reported no cases of diphtheria over the preceding 15 years and found that 37% of the children were carriers of "avirulent" diphtheria bacilli. And finally, in a careful epidemiological study carried out at the Royal Naval College in Greenwich, Dudley (1923) isolated typical nontoxigenic "avirulent" diphtheria bacilli from the throats of several Schick-positive boys during the course of an epidemic. From one individual who developed diphtheria, a toxigenic organism was isolated that was indistinguishable in all observable characteristics from a nontoxigenic strain isolated from the same boy on four previous occasions.

> *Pour certains peuples primitifs, le nom qui désigne un être possède un valeur magique. Avoir connaissance du nom d'un être, c'est pénétrer sa nature intime.*
>
> *Pour certains bactériologistes, le nom d'une bactérie semble lui aussi doué d'une vertu miraculeuse.*
>
> A. LWOFF, *L'espèce bacteriènne,* 1958

These observations were most puzzling to the taxonomists, and many were of the opinion that, despite apparent similarities, if a corynebacterium failed to produce a toxin that was neutralizable by specific antitoxin, it was unworthy of inclusion among the species *C. diphtheriae.* The work of Freeman (1951), who first demonstrated the intimate relationship between lysogeny and toxinogeny soon confirmed and ex-

---

[1] Epidemics of mild sore throat have been reported in more recent years in which the etiological agent appeared to be a nontoxigenic *C. diphtheriae.* Some years ago we had occasion to follow a mild sore throat with fever of short duration and an atypical membrane in a Schick-negative medical student who had inadvertently swallowed a culture of the nontoxigenic, nonlysogenic C7(−) strain through an unplugged pipet. The organisms were recovered in almost pure culture from the membrane.

tended by Groman (1953) and by Barksdale and Pappenheimer (1954), settled this taxonomic problem once and for all, but at the same time raised a host of far more intriguing questions.

> *Lysogeny occupies a privileged position at the crossroads of normal and pathological heredity, of genes and of viruses.*
> A. LWOFF, *Bacteriological Reviews*, 1953

What is the mechanism of this lysogenic conversion to toxinogeny, and what information is brought into the host bacterial cell by the prophage genome? Does phage DNA provide a structural gene that codes directly for the toxin protein, or does it control the synthesis of some other factor that in some way induces the host bacteria to form toxin? Is toxin a viral or a bacterial protein? Before we can attempt to answer these questions, it will be necessary to review some of the known facts about diphtheria toxin production and its relationship to pathogenicity, and to restate what is known about the nature of the toxin protein and its mode of action.

> *Thus we really have the impression that prophage behaves as a bacterial gene in the sense that it plays its part in the molecular orchestra.*
> A. LWOFF, *Harvey Lecture*, 1955

It is clear that in order to cause typical diphtheria, a corynebacterium must produce a specific toxin that is synthesized only by cells lysogenic for a particular phage β or one of its mutants. Other temperate phages such as γ, even though serologically related to β, fail to convert to toxinogeny. But although only cells carrying the β-phage genome are capable of synthesizing toxin, the *yield* of toxin is determined by the metabolism of the bacterial host cell and is not related in any known manner to phage growth. A few years ago, Mrs. Laura Sachs isolated in our laboratory several bacterial mutants of reduced toxin-producing capacity from cultures of C7($-$) and C7(β) strains treated with nitrosoguanidine. The properties of three of these mutants are summarized in Table 1 and compared with the wild-type C7($-$) and C7(β) strains and with a PW8 strain. One of the mutants, C7(β)NG54,

TABLE 1. PROPERTIES OF CERTAIN DIPHTHERIAL STRAINS

Sensitive strain	...	C7(−)	C7m4(−)	C7n2(−)	...
Lysogenized with β	PW8 *	C7(β)	C7m4(β)	C7n2(β)	C7(β)NG54
Generation time, min at 37°C	160	55	80–90	180	120
Spontaneous release of plaques, plaques on C7(−) per 10^9 bacteria	<1	3×10^4	10^5	10^3	Not done
Release of plaques after UV, plaques per 10^9 bacteria	1.3×10^1	$5\text{–}10 \times 10^9$	$2\text{–}5 \times 10^9$	$5\text{–}10 \times 10^7$	$5\text{–}10 \times 10^7$
Porphyrin release per 10^9 bacteria, n moles per 10^9 bacteria	1.4	0.16	0.11	0.02	<0.003
Toxin production, MLD per 10^9 bacteria	800–1000	80–120	15–20	5–10	0

* PW8 gives *ca.* 10^3 plaques per 10^9 cells before irradiation and 10^7 plaques after ultraviolet on the permissive host, *C. belfanti*. All other plaque counts on C7(−).

did not appear to release toxin under any conditions whatsoever, even when injected intradermally into guinea pigs and rabbits. Nevertheless, the β phage released by this strain converted wild-type $C7(-)$ to a typical lysogenic, toxinogenic $C7(\beta)$. Clearly, then, both phage and bacterial genes are involved and must cooperate in order to establish pathogenicity.

Among the host factors involved in toxin production, the most striking is the well-known effect of iron. Only when the bacterial content of iron and iron-containing respiratory enzymes is decreasing, is any significant amount of toxin synthesized. High yields, such as are obtained from PW8 (see Table 1), are produced only by organisms that are rapidly becoming "anemic" (Pappenheimer, 1955). There exists a stoichiometric relationship between a reduction in the cytochrome content of the bacteria, secretion of coproporphyrin III (a precursor of protoheme), and extracellular toxin release. It was this relationship that prompted us, many years ago, to suggest that diphtheria toxin might be related structurally to a bacterial cytochrome, probably of the b type (Pappenheimer, 1947). This hypothesis was never proved. However, recent observations that toxin binds both oxidized and reduced forms of nicotinamide adenine dinucleotide (NAD) suggest that, even though it may not be an integral part of a cytochrome molecule, toxin may nonetheless be related to an enzyme in the electron transport system.

But the potential capacity of a given strain to produce high yields of toxin under laboratory conditions does not correlate well with the severity of the disease it is capable of causing, even in highly susceptible populations (Mueller, 1941). A classic example is furnished by the Park-Williams Number 8 strain, isolated in 1895 from a mild case of diphtheria. Despite its feeble pathogenicity, this strain is used commercially throughout the world because of the high yields of toxin it produces, which often exceed 5% of the total bacterial protein synthesized and account for virtually all of the extracellular protein (Hirai et al., 1966). Under identical conditions, most strains isolated from severe cases of epidemic diphtheria produce only 5–10% as much toxin. As Mueller (1941) pointed out, however, it is the amount of toxin a strain is capable of producing *in vivo* that determines its relative virulence. The low pathogenicity of PW8 is not difficult to explain.

This strain has a defective cytochrome system, and, unlike most strains isolated from patients with clinical disease, it must rely entirely on flavoprotein as a terminal oxidase for all substrates except succinate (Pappenheimer, Howland, and Miller, 1962). Because of this defect PW8 grows in air at only about one-third the rate of wild-type strains. For this reason it cannot become established at the high iron level and low oxygen tension existing in tissues.

Still other factors, concerned with "invasiveness" and as yet ill defined, must have been acquired by *C. diphtheriae* during its evolution as a human pathogen to enable it to establish itself in local tissue lesions, from which it releases its toxin into the blood stream (O'Meara, 1940; Burnet, 1953). For example, typical diphtheria in man is not caused by various strains of corynebacteria, such as *ulcerans, belfanti,* and *ovis,* that can readily be isolated from the nasopharynges of horses and sheep. Yet, as recently shown by Staniča et al. (1968), a significant proportion of these animals carry *C. ulcerans* and *C. ovis* that are lysogenic and produce diphtheria toxin (see also Henriksen and Grelland, 1952).

There is another interesting and puzzling peculiarity of the PW8 strain. Despite the high yields of toxin produced by PW8, it has been difficult to demonstrate that the strain carries a toxinogenic phage (Barksdale, 1959). After UV irradiation most of the cells lyse within about 1.5 generation times (as compared with 3 for most lysogenic strains of *C. diphtheriae*), but only about one particle capable of forming a plaque on C7($-$) is released for every 10^7 PW8 cells that lyse (Miller, Pappenheimer, and Doolittle, 1966). Nevertheless, examination of such lysates in the electron microscope reveals phages that are morphologically indistinguishable from β (Mathews, Miller, and Pappenheimer, 1966).

Recently, Maximescu (1968) has reported that PW8 cultures spontaneously release toxinogenic phages that give rise to plaques on certain sensitive strains of *C. ulcerans* and *C. belfanti*. Using a strain of *C. belfanti* kindly sent us by Dr. Maximescu, Carolyn Melton has found that almost every PW8 cell will burst after UV irradiation and give rise to a plaque-forming particle. The average burst size is less than 2, presumably because of the defective respiratory system and slow growth rate of PW8 strains. Mrs. Melton has shown that the C7($-$) strain, commonly used as an indicator strain for β since the time of Freeman, is a restrictive host that apparently destroys phage DNA from PW8 unless that DNA is modified in some as yet undetermined manner. She has found that typical β phage from C7(β), after growth in *C. belfanti,* no longer forms plaques on C7($-$). The finding of a sensitive indicator strain for the phage released by PW8 has enabled her to select "cured," nontoxinogenic PW8 mutants.

What is the nature of the toxin that is produced only by virus-infected cells of decreasing cytochrome content, and how does it exert its lethal action in susceptible animals? Diphtheria toxin is not difficult to isolate as a crystalline heat-labile protein of molecular weight *ca.* 65,000 (Pope and Stevens, 1958a; Kato et al., 1960; Raynaud, Bizzini, and Relyveld, 1965), but until recently little was known of its mode of action. I think the real breakthrough that finally led to the solution of this problem at the molecular level was made by Strauss and Hendee (1960), who found that low concentrations of toxin block amino acid incorporation into protein by HeLa cells growing in culture. When a "saturating dose" of toxin ($>10^{-8}$M) is added to a culture of HeLa cells, incorporation of labeled amino acids into cell protein continues at its normal rate for 1.5–2 hours and then comes to a complete standstill. Exposure to very much lower concentrations (as low as $1-2 \times 10^{-13}$M!) will arrest protein synthesis within 2 or 3 days (Lennox and Kaplan, 1957; Gabliks and Solotorofsky, 1962; Pappenheimer et al., 1965). While excess toxin blocks growth and protein synthesis irreversibly within 2 hours, cells in spinner culture may retain their normal morphology by electron microscopy, normal permeability barriers, and other metabolic activities for 20 hours or even longer.

Cell-free extracts from intoxicated HeLa cells are still capable of incorporating labeled amino acids into polypeptides, but at a rate that is far below that observed in similar extracts from normal cells. Moreover, when low concentrations of diphtheria toxin are added to cell-free extracts from normal mammalian cells (even from species that are highly resistant to toxin), amino acid incorporation is strongly inhibited. The inhibition of protein synthesis in cell-free extracts by toxin requires the presence of NAD as a specific cofactor (Collier and Pappenheimer, 1964). The requirement for NAD is highly specific, and it cannot be replaced even by such close analogues as NADH and NADP (Goor and Pappenheimer, 1967a).

The specific target site of toxin and NAD action was identified by Collier (1967), who showed that inhibition of protein synthesis in cell-free extracts is due to inactivation of transferase 2 (T2), one of the enzymes required for polypeptide chain elongation. As has been demonstrated by Skogerson and Moldave (1968), T2 in the presence of GTP effects the translocation of the growing peptide chain on the ribosome so as to be in position to accept the next amino acid in the chain. Goor and Pappenheimer (1967b), who independently concluded that toxin acted by inactivation of T2, found that, of all the numerous factors required for protein synthesis, soluble T2 was the *only* one missing from extracts of intoxicated cells. Thus it appeared likely that inhibition of protein synthesis, first observed by Strauss and Hendee (1960), was due to primary action of the toxin on a particular essential enzyme.

The danger of parachuting young enthusiastic scientists into a flowerbed of selected data and fully bloomed conceptions should not be underestimated. A. LWOFF, *Bacteriological Reviews*, 1953

Quantitative studies of the inhibition of protein synthesis in cell extracts revealed an inverse relationship between toxin and NAD over a wide range of concentrations, so that for 50% inactivation of T2 the product [toxin] $\times$ [NAD] remains constant (Goor, Pappenheimer, and Ames, 1967; Gill et al., 1969). This observation suggested that the reaction probably involved the formation of a toxin-NAD complex. In fact, the binding of NAD by toxin has been demonstrated by equilibrium dialysis (Goor and Pappenheimer, 1967a), and of NADH by fluorescence quenching (Montonaro and Sperti, 1967). Even more significantly, it was observed by Goor et al. (1967) that the inactivation of T2 by toxin and NAD could be reversed by addition of moderate concentrations of nicotinamide, provided NAD was first removed by dialysis or by addition of streptococcal NADase.

The reversal by nicotinamide, the NAD requirement, and the low concentrations of toxin needed (especially *in vivo*) should have suggested to us earlier that T2 was being inactivated catalytically. This has now been shown, independently by Honjo et al. (1968) and by Gill et al. (1969), to be the case. Diphtheria toxin turns out to be an enzyme of a unique type that catalyzes the splitting of NAD with transfer of its ADPR moiety to be linked covalently, presumably at a point at or near the active site of T2, according to the reaction:

$$NAD + \text{transferase } 2 \rightleftharpoons \text{ADPR-transferase } 2 + \text{nicotinamide}$$

No protein other than free T2 itself has been found which can act as an acceptor for the ADP-ribose moiety. Even T2 that is bound to ribosomes can no longer react with NAD; in fact, as shown by Gill, Pappenheimer, and Baseman (1969), there is a competition between ribosomes and toxin for "free" T2.

By the technique of radioautography, it has been shown that even in the presence of a saturating dose of ^{125}I-labeled toxin a maximum of 25–50 molecules is bound per human HeLa cell (Pappenheimer and Brown, 1968). These few toxin molecules, located at the cell membrane, suffice to inactivate all the free T2 within the cell in the space of 10–20 minutes. Can such a high turnover number *in vivo* be accounted for by the activity of toxin observed *in vitro*? The cell concentration of toxin calculated from the above radioautographic observations is about 2×10^{-11}M, and the intracellular NAD concentration is about 5×10^{-4}M. The product, 10^{-14}M^2, agrees remarkably well with the [NAD] $\times$ [toxin] product of 3×10^{-14}M^2 found for 50% inhibition of amino acid incorporation by toxin in cell extracts. Finally, that the arrest of protein synthesis *in vivo* is due to the formation of inactive ADP-ribosylated T2 in the living HeLa cell is substantiated by the finding (Gill et al., 1969) that the inactive T2 in dialyzed ribosome-free preparations from extracts of intoxicated cells can be restored to full activity by addition of *both* nicotinamide and diphtheria toxin. Thus the very agent that caused intoxication in the first place is required for reversal of the inactivation *in vitro*.

We conclude that, regardless of the complexities that will surely arise when we attempt to explain the action of diphtheria toxin in animals, its primary action at the molecular level is to catalyze the transfer of the ADP-ribose moiety from NAD to T2 with the liberation of free nicotinamide and formation of an inactive enzyme. Preliminary experiments with organs and tissues taken from intoxicated guinea pigs and rabbits have shown a striking correlation between decreased amino acid incorporation and T2 inactivation (Baseman, Pappenheimer, and Gill, 1970).

A recent finding by Collier and his colleagues (Collier and Cole, 1969; Collier and Traugh, 1969), taken together with certain observations made in our own laboratory, may lead to a final solution of the puzzling problem of why the presence of a specific prophage is essential for toxinogeny. While purifying a toxin preparation by gel filtration, Collier was surprised to find that a substantial fraction of its T2-ADP-ribosylating activity was retarded and emerged from the column *behind* the major protein peak normally expected to contain the toxin. Further investigation revealed that treatment of purified toxin with the reducing agent dithiothreitol resulted in the breakdown of the 4.2S toxin molecule into smaller subunits. Although the specific T2-ADP-ribosylating activity of the subunits appeared to be even greater than that of the intact, native toxin molecules, virtually all of the original toxicity for HeLa cells or for guinea pigs had been lost. Some years earlier Pope and Stevens (1958b) had shown that crystalline toxin which had been subjected to treatment either with strong alkali or with sodium sulfite in alkaline solution gave three bands on agar diffusion against diphtheria antitoxin, all of which formed lines of immunologic identity with untreated toxin. Raynaud, Bizzini, and Relyveld (1965) confirmed and extended the findings of the British workers and interpreted them as signifying that diphtheria toxin is composed of at least two dissimilar subunits, each carrying different antigenic determinants.

In our laboratory (Gill, unpublished), we have found that, whereas a 4-times recrystallized preparation of diphtheria toxin sent us by Dr. M. Yoneda moved as a single sharp band corresponding to MW = 65,000 on polyacrylamide gel electrophoresis in the presence of sodium dodecylsulfate (SDS), after short treatment with a reducing agent such as β-mercaptoethanol or dithiothreitol it was almost completely broken down into two *unlike* subunits with molecular weights approximating 40,000 and 25,000. It seems clear from Collier's work that, although subunits are capable of catalyzing the transfer of ADPR from NAD to T2 *in vitro,* intact undissociated molecules are needed for T2 inactivation in living cells. Presumably, only one of the subunits carries the enzymatic activity of the toxin molecule, the other being needed for attachment to the cell membrane.

One might suppose that, once a toxin molecule has been attached to a sensitive cell membrane, it becomes split with its T2-inactivating subunit, entering the interior of the cell while the other fragment remains fixed at the cell periphery. Such a mechanism seems most unlikely, however, from autoradiographic studies of cells intoxicated with ^{125}I-labeled toxin (Pappenheimer and Brown, 1968), in which most if not all of the label was seen located in the cell membrane. Since almost no ^{125}I could be detected inside the cell, it would be necessary to postulate that the active fragment was not labeled with iodine. This was clearly not the case, since polyacrylamide gel-SDS electrophoresis of reduced toxin trace-labeled with ^{125}I showed that the label was approximately evenly distributed between the two subunits.

> *Les virus n'ont pas manqué de suivre la loi commune. Ce sont des parasites strictes qui, nés du désordre, ont créé, pour assurer leur perpétuation, un ordre nouveau des plus remarquables.*
> A. LWOFF, *Les Prix Nobels*, 1965

In our laboratory, Mrs. Carolyn Melton is presently studying the proteins released by a "cured" nonlysogenic PW8 strain of *C. diphtheriae,* growing under conditions of iron deficiency. She has found that under these circumstances this strain releases no toxin and only traces of ADP-ribosylating activity. The culture filtrates do, however, contain considerable amounts of a protein which, even without reduction or heat, moves with almost the same mobility on SDS-polyacrylamide gel electrophoresis as does the 25,000 dalton subunit of toxin. It is too soon to say more about the possible relationship between this extracellular protein released by the nontoxinogenic PW8 strain and the toxin subunit. Nevertheless, we have been tempted to take as our working hypothesis the intriguing possibility that toxin may be a genetic hybrid molecule with synthesis of one of its subunits determined by the phage genome and the other by the genome of the host bacteria.

CONCLUSIONS

We know very little about how it is that certain bacteria such as the diphtheria bacillus have become adapted so as to enjoy a happy parasitic existence in the human throat and nasopharynx, whereas other species—for example, *Aerobacter, Bacillus subtilus,* and *Escherichia coli*—fail to survive in the same environment. For an organism whose normal habitat is the throat to cause serious infectious disease in man it must have acquired still other specialized qualities. In the present instance we now know something about the complex events that must have taken place before the diphtheria bacillus became equipped to cause epidemic diphtheria, but we can only guess as to the order in which these evolutionary events may have arisen.

Nontoxinogenic, nonlysogenic *C. diphtheriae* are found in the nasopharynges of many domestic animals, as well as in the nasopharynx of man. At present there is no way for us to tell whether man or an animal species was the original host or reservoir for the diphtheria bacillus. The mere adaptation to survive in the nasopharynx of man probably required neither infection with a temperate phage nor the capacity to produce a lethal toxin. However, it is easy to see how an adaptation enabling the organism to grow and synthesize protein for short periods of time in the absence of an exogenous iron supply may have had survival value. In any case, this property must have been acquired before infection with a toxinogenic phage could effect toxin synthesis.

We have seen that diphtheria toxin is composed of two dissimilar subunits. Only one of the subunits possesses enzymatic activity and catalyzes the splitting of NAD with ADP ribosylation and consequent inactivation of the translocase required by mammalian cells for polypeptide chain elongation. Presumably, it is this subunit that is related to the protein moiety of an NAD-linked iron-dependent bacterial enzyme. The other subunit, under viral control, would therefore be required for attachment to the mammalian cell membrane. If this speculation should prove to be correct it will be surprising; but, then, most evolutionary events are the result of unexpected and accidental "happenings."

ADDENDUM ADDED IN PROOF

This essay was written one year ago and since that time new facts have emerged. It has been established that diphtheria toxin is synthesized and released as a single enzymatically *inactive* polypeptide chain of 62,500 daltons. Only after *both* a disulfide linkage and a peptide bond have been split is a fragment (not a subunit) of 24,000 daltons released that catalyzes the ADP ribosylation and inactivation of T2 (Gill and Dinius, Gill and Pappenheimer, Drazin and Collier, *J. Biol. Chem.*, in press). It now seems most probable that this activation takes place at the animal cell membrane so that one or more of the 24,000 dalton fragments reach the cell interior leaving 38,000 fragments fixed to the cell membrane (Baseman, Pappenheimer, and Gill (1970), *J. Exp. Med.*, **132**, 1138).

REFERENCES

Andrewes, F. M., ed. (1923). *Diphtheria,* London, His Majesty's Stationery Office.

Barksdale, W. L. (1959). *Bact. Rev.,* **23**, 202.

Barksdale, W. L., and A. M. Pappenheimer, Jr. (1954). *J. Bact.,* **67**, 220.

Burnet, F. M. (1953). *The Natural History of Infectious Disease,* Cambridge, Cambridge University Press.

Collier, R. J. (1967). *J. Mol. Biol.,* **25**, 83.

Collier, R. J., and H. A. Cole (1969). *Science,* **164**, 1179.

Collier, R. J., and A. M. Pappenheimer, Jr. (1964). *J. Exp. Med.,* **120**, 1019.

Collier, R. J., and B. A. Traugh (1969). *Cold Spring Harbor Symp. Protein Synthesis,* **34**, 589.

Dudley, S. F. (1923). *The Schick Test, Diphtheria and Scarlet Fever,* Med. Res. Council (Brit.) Spec. Rep. Series No. 75, London, His Majesty's Stationery Office.

Freeman, V. J. (1951). *J. Bact.,* **61**, 675.

Gabliks, J., and M. Solotorofsky (1962). *J. Immunol.,* **88**, 505.

Gill, D. M., A. M. Pappenheimer, Jr., and J. B. Baseman (1969). *Cold Spring Harbor Symp. Protein Synthesis,* **34**, 595.

Gill, D. M., A. M. Pappenheimer, Jr., R. Brown, and J. T. Kurnick (1969). *J. Exp. Med.,* **129**, 1.

Goor, R. S., and A. M. Pappenheimer, Jr. (1967a). *J. Exp. Med.,* **126**, 913.

Goor, R. S., and A. M. Pappenheimer, Jr. (1967b). *J. Exp. Med.,* **126**, 899.

Goor, R. S., A. M. Pappenheimer, Jr., and E. Ames (1967). *J. Exp. Med.,* **126**, 923.

Groman, N. B. (1953). *J. Bact.,* **66**, 184.

Henriksen, S. D., and R. Grelland (1952). *J. Path. Bact.,* **xiv**, 503.

Hirai, T., T. Uchida, Y. Shinmen, and M. Yoneda (1966). *Biken's J.,* **9**, 19.

Honjo, T., Y. Nishizuka, O. Hayaishi, and I. Kato (1968). *J. Biol. Chem.,* **243**, 3553.

Kato, I., H. Nakamura, T. Uchida, J. Koyama, and T. Katsura (1960). *Jap. J. Exp. Med.,* **30**, 129.

Lennox, E. S., and A. S. Kaplan (1957). *Proc. Soc. Exp. Biol. Med.,* **95**, 700.

Loeffler, F. (1887). *Centralbl. Bakt.,* 2, 105.

Lwoff, A., and A. Gutmann (1950). *Ann. Inst. Pasteur,* 78, 711.

Lwoff, A., L. Siminovitch, and N. Kjeldgaard (1950). *Ann. Inst. Pasteur,* 79, 815.

Mathews, M., P. A. Miller, and A. M. Pappenheimer, Jr. (1966). *Virology,* 29, 402.

Maximescu, P. (1968). *J. Gen. Microbiol.,* 53, 125.

Miller, P. A., A. M. Pappenheimer, Jr., and W. F. Doolittle (1966). *Virology,* 29, 410.

Montonaro, L., and S. Sperti (1967). *Biochem. J.,* 105, 635.

Moss, W. L., C. G. Guthrie, and J. Gelien (1921). *Johns Hopkins Hosp. Bull.,* 32, 109.

Mueller, J. H. (1941). *J. Immunol.,* 42, 353.

O'Meara, R. A. Q. (1940). *J. Path. Bact.,* 51, 317.

Pappenheimer, A. M., Jr. (1947). *Fed. Proc.,* 6, 479.

Pappenheimer, A. M., Jr. (1955). *Symp. Soc. Gen. Microbiol.,* 5, 40.

Pappenheimer, A. M., Jr., and R. Brown (1968). *J. Exp. Med.,* 127, 1073.

Pappenheimer, A. M., Jr., R. J. Collier, and P. A. Miller (1963). In *Cell Culture in the Study of Bacterial Disease,* edited by M. Solotorofsky, New Brunswick, Rutgers University Press.

Pappenheimer, A. M., Jr., J. H. Howland, and P. A. Miller (1962). *Biochim. Biophys. Acta,* 64, 229.

Pope, C. G., and M. F. Stevens (1958a). *Brit. J. Exp. Path.,* 39, 139.

Pope, C. G., and M. F. Stevens (1958b). *Brit. J. Exp. Path.,* 39, 150.

Raynaud, M. (1965). *Proceedings of the Federation of European Biochemical Societies,* vol. 1, p. 197, London, Pergamon Press.

Raynaud, M., B. Bizzini, and E. H. Relyveld (1965). *Bull. Soc. Chim. Biol.,* 47, 261.

Roux, E., and A. Yersin (1888). *Ann. Inst. Pasteur,* 2, 629.

Skogerson, L., and K. Moldave (1968). *Arch. Biochem. Biophys.,* 125, 497.

Stanica, E., P. Maximescu, C. Stoian, A. Pop, R. Oprison, and E. Potorac (1968). *Arch. Roum. Path. Exp. Microbiol.,* 27, 555.

Strauss, N., and E. D. Hendee (1960). *J. Exp. Med.,* 109, 145.

Herpesviruses, man, and cancer

or

The persistence of the viruses of love

◈

PROLOGUE

Herpesviruses share in common two unique distinctions. The first is that in the animal kingdom they are the only viruses of love. In man, one herpesvirus is transmitted by kissing (Hoagland, 1952; Evans, 1960), another by sexual intercourse (Nahmias and Dowdle, 1968) and a third by wrestling (Selling and Kibrick, 1964; Wheeler and Cabaniss, 1965), which, as is well known, may also be a manifestation of love. Nor does man suffer alone. At least some herpesviruses of cattle and horses are transmitted by sexual contact (Bowling et al., 1969); and, while transmission by kissing in animals has not been adequately investigated, at least one king, several virologists, and numerous common folk have learned too late that the bite of a playful monkey packs a lethal herpesvirus. The remarkable association of herpesviruses with love was clearly hinted at by Shakespeare in the much quoted lines by Mercutio in Romeo and Juliet:

> O'er ladies' lips who straight on kisses dream
> Which oft the angry Mab with blisters plagues
> Because their breaths with sweetmeats tainted are.

The association occurs not only in culture but also in the laboratory. Thus cells, which in culture often act as if they hold each other in utmost contempt, upon infection act as if they had ingested a potion of love in that they form clumps or large aggregates and even fuse (Roizman, 1962).

The second unique characteristic of herpesviruses is that they tend to persist, following primary infection, for the life of the organism (Burnet and Williams, 1939). Most herpesviruses studied to date appear to have the capacity to persist in their natural host. The presence of the virus is not usually apparent, but in some individuals lesions periodically occur from which the virus is readily isolated. In man the lesions recur most visibly around the lips, most painfully on the cornea, most dangerously in the central nervous system, and, alas, the least suspected on the cervix. The extraordinary and puzzling feature of herpesvirus infections is not that they recur, but rather that the recurrences are triggered in a predictable fashion in individuals showing respectable amounts of antibody to the virus. It is this puzzle which is the topic of this essay. There remains to be said how I came to be interested in it.

Chronologically this study began in 1957. This was one year after I received my degree and while I was working (happily, I might add) on the immunochemistry of poliovirus in a blacked-out room—one of the myriad now euphemistically known as The Johns Hopkins Medical Institutions. During that year Dr. Thomas B. Turner was made dean and I inherited his graduate student, one Malcolm David Hoggan. David, to my chagrin, was already working on the multiplication of herpes simplex virus. The work, he kept telling me, was going well; and, having failed with both carrot and stick to move him into the poliovirus fold, I began to read about herpesviruses. The significance of the problems I discovered for myself seemed clear, and like many others before me I began to think, rather vaguely, of lysogeny as a model for the persistence of herpesviruses in man.

It was not until 1958, when a Squibb Lectureship brought André Lwoff to Hopkins, that the aluminum foil lining the windows of the laboratory was finally removed (one of my more imaginative colleagues gave up cooking cells with neutral red and visible light), and I literally and figuratively began to see the light of day. That was a most decisive moment, for I knew exactly what I had to do next. To pave the way, however, took some time. Finally, at about 7 A.M. on September 19, 1961, I eagerly presented myself at the locked doors of André Lwoff's laboratory for what was undoubtedly the most exciting and profitable year of my life. Subsequently by mutual consent we began our working day a little later than 7 A.M. There was, in fact, an unspoken understanding that to beat each other to the laboratory by just a few minutes was quite sporting; too many minutes was not. That, however, is another story; this one belongs to herpesviruses.

On my return from Lwoff's laboratory I began working on the biochemical aspects of herpes simplex virus structure and multiplication, and with the exception of one brief fling (Roizman, 1965) I gave up what ambition I might have had for solving single-handedly the puzzle of how the virus persists in man. It seems, however, very fitting that in this book dedicated to André Lwoff I should review some of the problems, puzzles, and challenges that brought me to him in the first place. As a concession to the years that went by, I shall first construct an image of recurrent herpetic eruptions as seen from the point of view of molecular virology today and later attempt to determine just how well this image fits the available data concerning the growing number of herpesviruses associated with tumors of man and animals.

For the sake of clarity, it seems desirable to point out that the paper has a thesis, which may be summarized as follows. (i) After primary infection, herpes simplex virus is harbored in man in cells associated with sensory nerve trunks. (ii) In cells harboring the virus, the multiplication is arrested after the virion is uncoated, but before the early functions of the virus leading to the inhibition of the host are expressed. (iii) Suppression of viral macromolecular synthesis in the cells harboring the virus may be initiated and terminated by the nerve trunk or by nerve endings controlling these cells. (iv) A number of herpesviruses have been described which are associated with tumor cells in man and animals. The relationship of these viruses with the tumor cells is reminiscent of the relationship between herpes simplex and the cell harboring it. In this instance, the cell harboring the tumor-associated herpesvirus is the tumor cell. Moreover, the mechanisms which regulate viral macromolecular synthesis in these cells may be different.

THE PROBLEMS

Man becomes infected with the labial (Subtype 1) variants of herpes simplex virus in childhood, with the genital (Subtype 2) around the age of consent. Most people recover from primary infection without serious consequences, but some individuals are afflicted with recurrent herpetic eruptions of the lip, cornea, genitals, and other parts of the body for the rest of their lives. What makes this recurrent infection unique are three facts. First, the individuals suffering from recurrent eruptions almost invariably have circulating antibody. Second, new recurring lesions at different sites may be established by natural or artificial inoculation of autologous or entirely different virus strains (Terni and Roizman, 1970; Nahmias and Dowdle, 1968; Nahmias, Chiang, del Buono, and Duffey, 1969). Lastly, patients suffering from recurrent infection can often predict the recrudescences accurately; they follow a specific physical or emotional provocation, such as fever resulting from certain bacterial and parasitic infections, menstruation, intake of "the pill," cortisone, or exposure to sunlight and wind (Terni, 1965; Roizman, 1965; Morgan, 1968; Buddingh et al., 1953; Scott, 1957). The eruptions can be both induced and suppressed by psychotherapy and surgical intervention (Blank and Brody, 1950; Carton and Kilbourne, 1952; Carton, 1953; Cushing, 1905; Heilig and Hoff, 1928; Schneck, 1944).

These observations constitute epidemiologic findings so unlike those of other viral diseases as to have misled Doerr (1938) to postulate that the recrudescences are caused by a virus generated each time *de novo*. It is now generally accepted, however, that after primary infection the virus is harbored in an inapparent form at some particular site. Specific stimuli associated with physical and emotional provocations of the host cause the virus to manifest itself in the form of recurrent eruptions. The problems concerning us are the site and mechanism of survival of the virus, and the mechanism by which the eruption is activated.

ANALYSIS OF RECURRENT ERUPTIONS

THE SOURCE OF THE VIRUS RESPONSIBLE
FOR THE RECURRENT ERUPTIONS

The fact that individuals suffering from recurrent eruptions can accurately predict the recrudescences argues strongly in favor of the hypothesis that the virus persists at some site in the body in the interim between recrudescences. But where?

Topologically, herpetic eruptions generally recur on approximately the same site of the body. The question thus arises whether the virus persists at the site of the recurrent eruption or at some distant site. Despite the isolation of virus from tears and saliva of individuals with recurrent infections (Kaufmann, Brown, and Maloney, 1967), the evidence on the whole favors the hypothesis that the virus survives in the interim between recrudescences at or very near the site of recurrent eruptions. Briefly, this evidence is as follows. (i) Removal of the afflicted skin has been reported to suppress recurrent eruptions for varying intervals. Eruptions continue, however, to recur on adjacent areas and ultimately also on the area of the excised skin (Nicolau and Poincloux, 1928; Stalder and Zurukzoglu, 1936; Antonelli and Vignali, 1968). (ii) It is relatively easy to establish recurrent infections at new sites distal from lacrimal and salivary glands. Inadvertently, this is often done by physicians who attempt to exorcise the virus by vaccinating the patient with fluid aspirated from the recurring vesicles. Frequently the net consequence is a recurrent eruption at the site of vaccination as well as at the old site. (iii) Simultaneous recurring eruptions above and below the waist have also been reported (Terni and Roizman, 1970; Nahmias and Dowdle, 1968). The significance of this finding stems from the fact that eruptions occurring at different sites manifest only the virus characteristic for that area of the body; the viruses responsible for the eruptions at each site do not circulate freely throughout the body.

The exact nature of the cell which harbors the virus in the interim between recrudescences is unknown, but the finger of suspicion has long pointed to nerve cells and to cells associated with sensory nerve fibers (Goodpasture, 1929; Paine, 1964). The evidence is circumstantial and based on the following: (i) In primary infection herpesviruses may readily gain access to the central nervous system (Goodpasture, 1929). Herpes simplex virus has been associated with severe encephalidities, and encephalitis-like syndromes have also been described in patients with recurrent herpetic lesions (Warren, Carpenter, and Boak, 1940; Leider et al., 1965). (ii) The distribution of vesicles in recurrent herpes infection corresponds closely to the area of innervation of cutaneous nerves (Blank and Rake, 1955). Experiments *in vitro* show that the virus is able to infect nerve cells. However, the nerve cells are destroyed by the infection (Feldman, Sheppard, and Bornstein, 1968). Since in man sensory nerve cells do not appear to be affected, it would follow that the virus is associated with satellite cells rather than nerve cells themselves. (iii) Herpetic vesicles cannot be induced in denervated skin (Carton, 1953; Ellison, Carton, and Rose, 1959). (iv) The delay in the reappearance of recurrent herpetic eruptions following surgical removal of the affected skin suggests that the virus is maintained, not in the skin, but in some deeper tissue and is compatible with the hypothesis that the virus is transmitted via nerve trunks which must regenerate before eruptions reappear (Stalder and Zurukzoglu, 1936; Antonelli and Vignali, 1968). (v) Sectioning of the trigeminal nerve for alleviation of trigeminal neuralgias almost invariably causes an outcrop of herpetic vesicles along the distal portion of the sectioned nerve trunk (Carton and Kilbourne, 1952; Carton, 1953; Cushing, 1905).

THE STATE OF THE VIRUS IN THE INTERIM
BETWEEN RECURRENT ERUPTIONS

Numerous attempts have been made to isolate infectious virus from the site of recurrent eruptions in the interim between recrudescences; a few very extensive and thorough experiments have been reported (Corriell, 1963; Falchi, 1925; Findloy and MacCallum, 1940; Rustigan et al., 1966; Antonelli and Vignali, 1968). Without exception, all attempts to isolate virus from the tissue directly or by growing the cells *in vitro* first have been unsuccessful. These studies indicate that the virus is not actively multiplying at the site of recurrent eruptions in the interim between recrudescences and raise numerous questions concerning the state and location of the virus in this interim. A few deductions based on currently available information can be made:

1. The fact that the recurrent lesions are usually localized in a circumscribed area of the face, genitals, etc., suggests that the virus does not circulate freely in the blood. This hypothesis finds solace in the fact that people exhibiting current eruptions almost invariably have neutralizing antibody in their serum. It must necessarily be concluded that in the interim between recurrences the virus persists in some cells and does not circulate freely through the body.

2. It is most unlikely that the virus persists as an infectious, mature virion in the cell in which it has multiplied. This conclusion is based on relatively recent findings that the herpesvirion consists of a core, a three-layered capsid, and two envelopes (Roizman, Spring, and Schwartz, 1969). The inner envelope is acquired in the nucleus. The labial variants acquire the outer envelope as the particles bud through the modified inner lamella of the nuclear membrane (Fig. 1) (Darlington and Moss, 1968; Schwartz and Roizman, 1969a). The inner envelope is both sufficient to render the nucleocapsid infectious and necessary for the acquisition of the outer envelope. However, particles lacking the outer envelope are unstable. The envelopes collectively consist of host (preformed) lipids and several glycoproteins specified by the virus and synthesized in the cytoplasm of the infected cells. The glycoproteins vary considerably, depending on the virus strain (Keller, Spear, and Roizman, 1970; Roizman, 1970).

The pertinent property of the glycoproteins is that, in addition to binding to the nuclear membrane which ultimately becomes the outer envelope, they also modify the endoplasmic reticulum (which then serves as the means of egress of virus from the cell; Schwartz and Roizman, 1969b) and the cytoplasmic membrane (Spear, Keller, and Roizman, 1970). The consequences of the binding of the glycoproteins to the cytoplasmic membrane are manifest in a change in the interaction with adjacent, uninfected cells and in a profound change in the immunologic specificity of the membrane surface (Roizman, 1962, 1969; Roane and Roizman, 1964; Roizman and Spring, 1967). It is largely on the basis of the drastic change in the surface antigenicity that we must conclude that a cell which produces infectious virus and by necessity, viral glycoproteins, cannot possibly survive very long in an immune host.

3. There is a more attractive alternative, namely, that the persistence of the virus in the interim between recrudescences is dependent

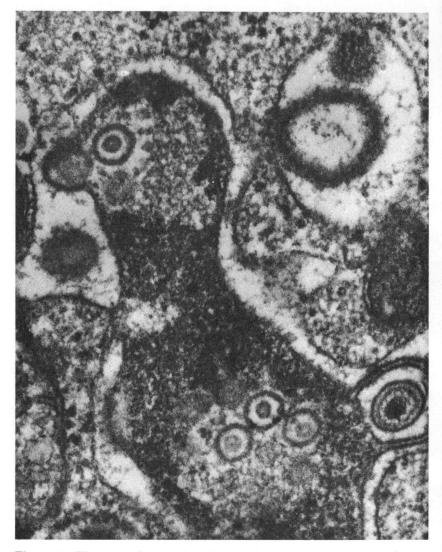

Figure 1. Electron microscopic evidence for the envelopment of herpes simplex virus at the inner lamella of the nuclear membrane. Virus particles are seen projecting through the inner lamella of the nuclear membrane into the space between the lamellae of the distorted nuclear membrane. The photomicrograph has previously been used to illustrate the extravagant delusions of electron microscopists. From Spear and Roizman, unpublished studies.

on the arrest of virus multiplication in the infected cell at some point
before the synthesis of glycoproteins specified by the virus. One obvious
question is whether there are any data which would permit us to determine at what point in the reproductive cycle the multiplication is
arrested.

One difficulty in pursuing this analysis is that most of our information comes from studies of herpes simplex virus in cells grown *in vitro.*
We must necessarily assume that the virus specifies the same structural
components in cells growing *in vitro* and *in vivo* and that these components will necessarily behave in the same way in both systems to produce infectious virus. It is not at all necessary, however, to assume that
the virus specifies the same nonstructural products or that they function
in an identical manner in the two systems.

Two specific points should be made. The first point is that the
cells grown in cell culture do not resemble either the cells in which the
virus is harbored or, in fact, the corresponding somatic cells from which
they are derived. Parenthetically, it should be pointed out that there is
a difference in the function of animal cells in the artificial *in vitro* environment of the cell culture and in the whole animal. In culture, cells act
as single entities competing independently for survival. In the animal,
they are dependent components of a complex multicellular organism
and readily expendable if required for the survival of the organism. The
second point concerns the virus. Simple calculations show that the virus
carries sufficient genetic information for nearly 150 proteins, 50,000
daltons each. Present data indicate that no more than 10% of this genetic
information specifies structural components of the virion (Spear and
Roizman, 1968; Roizman, 1969). The function of the products specified
by the rest of the genetic information is not clear. In view of the complex behavior of the virus *in vivo* and the rather simplistic lytic cycle in
cells cultivated *in vitro,* it cannot be excluded that the virus expresses
more genetic information *in vivo* than *in vitro.*

Bearing in mind this exposé of the inadequacy of data obtained
from studies of infected cells grown *in vitro* as they relate to human
infection, it is nevertheless desirable to extrapolate from them. Briefly,
two points can be made:

1. Most cell cultures, studied in our laboratory, when exposed to a herpes simplex virus strain, act as if it is the inherent property of the cell to attempt to express the genetic information contained in the virus with which it becomes infected. Particularly pertinent to this discussion are the observations that the entry and uncoating of the virus take place in both permissive and nonpermissive cells and that the uncoating of the viral DNA does not require the postentry synthesis of products specified either by the virus (as in the case of poxvirus infection) or by the cell (Roizman, 1969b). Moreover, the reproductive cycle of the virus follows a rigid, reproducible time table which is difficult to interrupt without starving the cell or using exotic drugs (Tankersley, 1964; Roizman, 1963). This is perhaps the appropriate place to point out that attempts to establish model systems of recurrent herpetic infection *in vitro* have generally failed (Roizman, 1965; Terni, 1965). The achievements, so far, have been virus-carrier cultures in which infectious virus can be demonstrated with ease. One necessary conclusion resulting from these observations is that, if the cells harboring the virus bear any resemblance to the cells *in vitro,* the arrest of virus multiplication occurs after uncoating of the DNA and that this interruption in multiplication is not a built-in, invariant feature of the reproductive program of the virus (Roizman, 1965).

2. A more precise estimate of the probable event at which the reproductive cycle is interrupted may be deduced from an analysis of cellular functions after infection. Particularly pertinent is the following: (*a*) Within a few hours after infection host chromosomes become aggregated and displaced to the periphery of the nucleus. Host DNA synthesis ceases before the initiation of synthesis of viral DNA (Ben-Porat and Kaplan, 1965; Roizman and Roane, 1964). (*b*) The synthesis of host ribosomal precursor RNA, heterodisperse RNA, and 4S RNA begins to decline immediately after infection and reaches 30% of control level at about 3 hours after infection (Wagner and Roizman, 1969). However, most of the RNA that is made is not processed. Thus, at 3 hours postinfection the 45S ribosomal precursor RNA that is made becomes methylated but is not processed into 18S and 28S ribosomal RNA. Coincidentally with the decrease in the synthesis and processing of RNA, the nucleolus becomes disaggregated (Love, Rabson, and Wildy, 1964)

and, so far as we can tell from autoradiographic studies (Schwartz and Roizman, unpublished data), ceases to function. (*c*) Immediately after infection host polyribosomes begin to disaggregate, host protein synthesis ceases, and, coincidentally, host glycoproteins no longer become incorporated into cellular membranes (Sydiskis and Roizman, 1967; Spear and Roizman, 1968; Spear, Keller, and Roizman, 1970). The inhibition of host macromolecular synthesis appears to be an essential prerequisite for the synthesis of structural components of the virus (Aurelian and Roizman, 1965). Moreover, the inhibition is so profound that it is unlikely that the cell can recover from infection; indeed numerous studies have failed to show survival of infected cells *in vitro*. We have no evidence that similar events also take place *in vivo*. However, electron photomicrographs of infected cells from acute herpetic infections of man and of experimental animals show very similar changes and strongly suggest that the infected cells are not likely to recover (Swanson, Craighead, and Reynolds, 1966; Rabin et al., 1968; Patrizi, Middelkamp, and Reed, 1968; Luse, Friedman, and Smith, 1965). It necessarily follows that in the cell harboring the virus in the interim between recrudescences the multiplication of the virus is arrested after its entry into the cell but before it expresses the functions leading to the inhibition of cell macromolecular biosynthesis and cell death.

CELLULAR DETERMINANTS OF VIRAL INFECTION

The chain of reasoning followed so far should lead us to believe (i) that the virus persists in cells in the interim between recrudescences and (ii) that in these cells the multiplication of the virus is reversibly arrested after the DNA is uncoated but before the virus expresses functions which lead irrevocably to the death of the cell. This conclusion, arrived at circuitously and only after many qualifications, has a very familiar ring to it (Lwoff, 1953). There remains to be determined whether the arrest of virus multiplication is primarily a viral function, as in lysogeny, or entirely a cellular function.

As pointed out earlier in this essay, we have no suitable *in vitro* model for analysis of this hypothetical mechanism for the persistence of the virus in man. Two points, however, should be made:

1. Herpesviruses are notoriously cantankerous and difficult to grow. There is now sufficient evidence that the virus multiplies best in rapidly

growing cells and only poorly, if at all, in cells maintained in a stationary phase (Roizman and Spear, 1968; Aurelian, 1968). Even more interesting is the observation that herpes simplex virus does not multiply in freshly harvested, untreated lymphocytes but multiplies readily in lymphocytes stimulated with phytohemagglutinin (Nahmias, Kibrick, and Rosan, 1964). These studies indicate that the virus requires host products which are involved in the multiplication of cells and which do not accumulate or persist in the absence of cell division.

2. If indeed the hypothetical arrest of viral multiplication is a cellular function in a manner analogous to that seen in lymphocyte suspensions *in vitro,* the question arises whether there exists a cell *in vivo* which is suitably distributed along sensory nerve fibers, can become infected, and exhibits at least some of the properties of the lymphocyte suspension maintained *in vitro*. We have already established that the cells harboring the virus are associated with nerve cells. Probably the most promising clue as to the properties of these cells is the behavior of the Schwann cells enveloping the peripheral stump of the nerve fiber undergoing Wallerian degeneration. Normally Schwann cells enveloping intact nerve cells do not multiply, and all functions associated with cell division appear to be inhibited (Causey, 1960). However, within several hours after sectioning of the nerve fiber, the Schwann cells along the nerve fiber peripheral to the section begin to multiply. Concomitantly, the DNA and RNA contents of these cells increase. The striking aspect of these observations is that trigeminal tractotomy, which apparently favors the multiplication of the virus, also favors the multiplication of Schwann cells. It could be predicted that, if the hypothetical arrest of viral multiplication is a cellular function, the macromolecular biosynthesis of cells harboring the virus is controlled by the nerve cells in the same fashion as that of the Schwann cells.

WHAT DETERMINES THE SITE OF LOCALIZATION OF RECURRENT ERUPTIONS?

As pointed out in the preceding sections, recurrent infections occur on different parts of the body but most frequently on the face, cornea, and genitals. Moreover, it has been demonstrated repeatedly that viruses associated with genital infections differ from those isolated from herpetic eruptions on other parts of the body with respect to numerous properties

(Dowdle et al., 1967; Ejercito, Kieff, and Roizman, 1968; Plummer, Waner, and Bowling, 1968; Goodheart, Plummer, and Waner, 1968; Figueroa and Rawls, 1969). These findings raise two important questions: first, what is the origin of these viral variants and, second, what determines the site of localization of the recurrent eruptions? It should be pointed out at the onset of this discussion that the experimental data bearing directly on these questions are very meager; however, the vague outlines of the answers are becoming apparent.

On the basis of neutralization tests, strains isolated from different parts of the body have been differentiated into two types: those isolated from genital lesions (Type 2) and those isolated from other parts of the body (Type 1) (Schneweiss, 1962; Dowdle et al., 1967; Ejercito et al., 1968). However, at this point several qualifying statements should be made. The first qualification is that, while it is relatively easy to differentiate between Type 1 and Type 2 strains with selected antisera, the two types are nevertheless immunologically related, and both can be neutralized by the same monotypic antiserum. The second point that should be made is that the tests usually employed to differentiate between Type 1 and Type 2 strains are not very sensitive. More sensitive neutralization tests employing an internal reference marker indicate that there may be considerable immunologic differences among freshly isolated strains within each type (Terni and Roizman, 1970). The notion that herpesviruses are highly variable in nature is also supported by studies of laboratory strains. Thus in the laboratory it has been rather easy to obtain mutants of established virus strains differing in immunologic specificity and in other physical and biologic properties (Roizman and Roane, 1963; Ejercito et al., 1968; Roizman and Aurelian, 1965). These findings have led to the proposal that the survival of the virus in the immune host is due to its variability in immunologic specificity and virulence (Ashe and Scherp, 1965; Hampar and Burroughs, 1969). (Alas, the explanation does not fit the facts, since the immunologic differences are not sufficiently large to permit the survival of the mutants in the immune host.) Very pertinent here is the finding that at least several Type 1 laboratory strains have produced mutants which are immunologically intermediate or clearly Type 2 strains (Dowdle et al., 1967; Ejercito et al., 1968). Recent analyses have revealed differences in the electrophoretic profiles of the glycoproteins contained in the envelopes of Type 1 and Type 2 viruses (Keller, Spear, and Roizman, 1970). As expected, immunologically intermediate strains produced in the laboratory showed profiles which were intermediates between those of Type 1 and Type 2.

It seems reasonable to conclude on the basis of this discussion that certain properties of herpesviruses are highly variable and that the genital strains were derived from nongenital strains and were selected because of a specific advantage in venereal transmission. The nature of this advantage is not known. Preliminary studies of the glycoproteins contained in viral envelopes suggest that the genital strains may have lost one or more of the glycoproteins (Keller, Spear, and Roizman, in preparation); the meaning of this finding is not clear. However, if localization of the eruptions caused by the genital strains is determined by the virus, it seems reasonable to ask (i) whether the localization of the recurrent lesions of the cornea, the thigh, and other parts of the body is also determined by some inheritable property of the virus, and (ii) just how the virus "determines" the site at which it is harbored.

With respect to the first question, the only available data are recent findings that virus isolates recovered from thigh infections of female patients differ in minor characteristics but reproducibly from prototype genital strains (Terni and Roizman, 1970). In view of the existence of virus strains which cannot be classified as either Type 1 or Type 2 (Dowdle et al., 1967; Ejercito et al., 1968), there is a distinct possibility that additional studies may in fact confirm and magnify further the differences among viruses isolated from recurrent eruptions of different organs of the body. With respect to the second question, that is, just how the virus is selected or how it determines the site of the recurrent infection, obviously the cells harboring the virus must at some time come into contact with the virus to become infected, and it must be compatible in some specific property with the virus they contain in order to survive. Beyond that we know nothing.

THE PRECIPITATION OF RECURRENT ERUPTIONS

In bygone years the appearance of a reddish spot heralding a recurrent eruption on the lip of a patient with acute pneumococcal pneumonia brought a sigh of relief from the attending physician; it meant that the prognosis was excellent and the patient would recover. Medical care, treatment, and the basis for arriving at a sound prognosis have changed, but fundamentally the recurrent eruptions are not ordinarily characteristic of well-being. In this section we shall examine briefly two questions: first, what the various physical and emotional factors which result in precipitation of recurrent eruptions have in common, and second, what information, if any, is available on the actual mechanism of precipitation of recurrent eruption.

A survey of the extensive literature on recurrent herpetic eruptions shows that the conditions and factors associated with recurrent infection may be divided into three groups. First, recurrent herpetic infections are more frequent and severe in patients suffering from malnutrition, leukemias, and certain types of hereditary immunologic deficiencies (St. Gene et al., 1962, 1965; Tokumaru, 1966). Second, it seems desirable to lump together individuals responding primarily to physical provocations such as pyrexia resulting from infection, artificially induced fever, sunlight, and wind, and, strangely enough, those responding to certain types of neurosurgery. Third, it also seems desirable to lump together individuals responding primarily to intense emotional provocations and those exhibiting recurrent eruptions in conjunction with menstruation and hormone treatment. It should be noted parenthetically that not all of the precipitating factors and conditions are equally effective. Trigeminal tractotomy is almost invariably effective in the selected population on which it is performed. Pyrexia comes next; in the age when syphilis was treated with artificial fever, 190 of 411 patients in one study developed herpetic eruptions following treatment (Warren, Carpenter, and Boak, 1940). Not all febrile diseases are equally efficient in precipitating eruptions. Herpetic eruptions are very frequent in malaria and in pneumococcal pneumonia; they are less frequent in brucellosis and very rare in typhoid fever. There is really very little information on the events and mechanisms that precipitate a herpetic eruption. The following points, however, should be noted.

1. For heuristic reasons it is desirable to make the assumption that the different kinds of cells capable of harboring the virus are limited in number and that each factor or condition triggering an eruption does not act on entirely different kinds of differentiated somatic cells.

2. It seems likely that trigeminal tractotomy and possibly fever, sunlight, and other physical provocations act directly upon the cell which

harbors the virus and not by inducing a factor which circulates through the body. The evidence is particularly clear in the case of recurrent eruptions resulting from trigeminal tractotomy. Thus the eruptions occur only in the areas innervated by the sectioned branch of the trigeminal nerve and not along the branches remaining intact. Patients suffering sensory loss as a result of partial tractotomy, on being operated upon again, frequently develop eruptions along the distal portion of the newly sectioned branches; no postoperative eruptions appear in the area of the skin anesthetized by the first operation. Clearly, in this instance the factor responsible for triggering the eruptions does not circulate in the blood.

3. It is also very likely that the eruptions associated with menstruation, hormone treatment, etc., are triggered by a factor circulating through the body of the patient. This is readily apparent in a recently described case in which menstruation triggered simultaneously recurring eruptions of the lip and thigh caused by different variants of the virus (Terni and Roizman, 1970). Simultaneously recurring eruptions on different parts of the body are not uncommon.

4. The common factor responsible for the increase in frequency and severity of eruptions in patients suffering from malnutrition, leukemias, and immunologic deficiency is not very clear. The possibility exists, however, that diseases characterized by debilitation and immunologic deficiency affect the termination rather than the initiation of herpetic eruptions. Specifically, it has been assumed that herpetic eruptions are terminated by immunologic mechanisms involving, among other events, massive invasion of leukocytes into the infected area. It is conceivable that the rate of response of the organism to the recurrent eruption varies considerably both in the same individual at different times and from person to person, and that in some people the eruptions may be terminated so soon after they began that the patient may not even become aware that a recrudescence was about to take place, whereas in others the immune response is slow and the healing is delayed. If this were true, two conclusions would necessarily follow. (*a*) Virus multiplication is triggered in more individuals than those suffering from recurrent eruptions. This would explain in large measure the finding that antibody to the virus tends to persist at relatively high levels in the population.

(*b*) The frequency and severity of recurrent eruptions may be a good indication of the immunologic competence and responsiveness of the individual.

HERPESVIRUSES AND CANCER

In rare instances recurrent herpes simplex is painful, interminable, and very dangerous. However, the disease is not a major medical problem. When this infection began to attract widespread attention, lysogeny was new, unknown in higher organisms, and relatively easy to define operationally. All this has changed. Bonding between host and adenovirus DNA has been reported (Doerfler, 1968), and consequently one operational definition of lysogeny has been demonstrated in eukaryotic cells, at least as a laboratory phenomenon. On the other hand it is clear that the lack of a suitable experimental model and the complexity of the multicellular organism make it most unlikely that covalent bonding between cellular DNA and herpes simplex DNA and immunity to reinfection with an identical virus, the two essential features of lysogeny, will ever be demonstrated. Moreover, lysogeny is not the only model uniquely suitable in this instance. In lysogenic bacteria, the survival of the phage genome depends on its ability to replicate at a rate comparable to or greater than that of the host. The survival of herpes simplex virus may not depend on its ability to multiply over very long intervals between recrudescences of the disease. This would be particularly true if the virus is harbored by cells controlled by nerve fibers which do not multiply. Why then the interest in recurrent herpetic infection?

One important discovery made in recent years is that there exist in nature numerous herpesviruses far more cantankerous and difficult to handle than herpes simplex and apparently associated with tumors of man, fowl, frogs, guinea pigs, rabbits, and monkeys (Epstein, Achong, and Barr, 1964; Wight et al., 1967; Hsiung and Kaplow, 1969; Melendez et al., 1969; Fawcett, 1956). The role of these herpesviruses in the causation of the tumors with which they become associated is not clear, but numerous features of this association are reminiscent of herpes simplex virus infection of man. Three brief examples should suffice.

One herpesvirus appears to be primarily associated with infectious mononucleosis (G. Henle, W. Henle, and Diehl, 1968), a disease of young adults which not infrequently recurs (Bender, 1962; Evans, 1963). The same or very similar agents are associated with lymphomatous tumors in young children in Africa and with postnasal carcinoma occurring in the elderly of Chinese extraction. The herpesvirus particles are not demonstrable in tumor cells *in vivo*. They are present, but in few cells and in small amounts, in tumor cells grown *in vitro*. The largest numbers of infected cells and the largest yields of virus particles are obtained from cells maintained *in vitro* under conditions of partial nutritional deprivation (Henle and Henle, 1968). However, regardless of the conditions of maintenance, the cells producing virus exhibit such obvious degenerative changes that it is most unlikely the cells are capable of surviving. Moreover, the virus produced *in vitro* is rarely transmitted as a cell-free extract.

The second example, Marek's disease, is an acute, usually fatal, highly infectious disease of poultry. The two interesting features of this infection are related to the transmission of the virus and to the tumorlike lesions which develop in a number of organs of the infected bird. Recent studies show that the virus probably responsible for the epidemics is transmissible in a cell-free filtrate and is made in the cells lining the feather follicles (K. Nazerian, personal communication). Virus particles are also present in other organs and in cell cultures made from these organs; however, the virus produced in these cells is not readily transmitted as a cell-free filtrate. Moreover, no virus particles are present in the tumor cells *in vivo*. The noteworthy characteristic of this infection is that, both *in vivo* and *in vitro,* the cells producing virus particles exhibit drastic degenerative changes and do not survive infection.

The relationship between the tumor and the herpesvirus associated with it is somewhat more apparent in the third example. The suspicion that adenocarcinoma of the frog is caused by a virus was first voiced by Lucké (1938). The presence of herpesvirus in the tumors was subsequently reported by Fawcett (1956) and others (e.g., Lunger, 1964). The epidemiology of the disease is not known; the virus made in culture is not transmissible as a cell-free filtrate. The pertinent property of the Lucké adenocarcinoma is that virus is present in the tumors only during the winter months or if the frog bearing the tumor is refrigerated for several weeks (Rafferty, 1964). Virus extracted from the tumors is infectious and a cell-free filtrate will produce tumors in tadpoles (Tweedell, 1967; Mizell, 1969). Virus particles are absent, however, from tumors of frogs collected during the summer months. Again characteristically, tumor cells induced to make virus exhibit degenerative changes characteristic of herpes-infected cells. While experimental data are lacking, it seems rather unlikely that they could survive once the reproductive cycle is completed.

The viruses found in African lymphoma, Marek's disease of poultry, and Lucké adenocarcinoma of the frog share in common three features with herpes simplex virus. First, the virus is not apparent in the cell harboring it without some specific manipulation of the cell. To manifest herpes simplex virus man must be physically or emotionally provoked. We do not know the nature of the stimulus which induces the herpesviruses to multiply in Burkitt's lymphoma or in the tumor cells of Marek's disease *in vivo;* to manifest the virus *in vitro* the cells must be grown and maintained under appropriate conditions. Finally, to demonstrate the virus in Lucké adenocarcinoma the frog must be refrigerated for several weeks. The second feature is the inhibition of host macromolecular metabolism. As we have seen, once the cell begins to make viral structural products and to assemble virions (even though they are not infectious!) the cell is doomed. The third feature in common and the one which may be most important is that the cell which harbors the virus may not be the one which optimally replicates the virus and maintains it in nature. In more precise terms, the cells harboring the virus are at best only *conditionally permissive.* This is clearly seen in Marek's disease and in recurrent herpes simplex; it is somewhat less clear in Burkitt lymphoma and in Lucké adenocarcinoma, in which the epidemiologically important mechanism of transmission is uncertain.

It should be noted finally that there is also a difference between herpes simplex and the tumor-associated herpesviruses. Whereas the tumor-associated herpesviruses are, by definition, harbored by tumor cells and may be responsible for the transformation, herpes simplex momentarily is free of such a taint. One wonders, however, for how long; already herpes simplex virus has been suspected as the causative agent of carcinoma of the cervix and of the squamous cell carcinoma occurring at the site of recurrent infections (Kvasnicka, 1963, 1964,

1965; Josey, Nahmias, and Naib, 1968; Rawls et al., 1968; Wyburn-Mason, 1957).

CONCLUSIONS

In large part this paper has dealt with the persistence of herpes simplex virus in man as a model system for the interaction of herpesviruses with their natural hosts. The observations cited herein indicate that the virus is harbored in man in the interim between recrudescences of the disease in cells associated with sensory nerve fibers. It was indicated that *in vitro* the virus multiplies best in rapidly growing cells. While it is dangerous to extrapolate from *in vitro* studies to *in vivo* conditions, there is in fact some basis for the hypothesis that nerve cells severely restrict both cellular and viral macromolecular metabolism in cells harboring the virus and that the restriction is alleviated and multiplication of the virus permitted when the nerve fibers are damaged or appropriately stimulated by fever, cold, hormones, etc. Since there is no evidence that the cells harboring the virus can grow in culture, the failure to isolate virus from cultures of skin afflicted with frequent herpetic eruptions does not detract from the basic hypothesis. Unfortunately, what really goes on is unknown.

We know considerably less about the herpesviruses associated with tumors than about herpes simplex virus. In some respects the information deduced from the model system is applicable to these viruses. It is clear that herpesviruses associated with tumors are harbored in specific cells which may vary from species to species and that, like herpes simplex, the viruses can be made to manifest themselves and destroy their host. This unity in behavior is not tarnished by apparent differences. Thus, whereas *in vitro* herpes simplex multiplies best in rapidly growing cells, the viruses associated with tumors appear to require a host debilitated by deprivation of an essential amino acid (Y. Becker, personal communication). Perhaps this too is a sign of unity. As mentioned earlier herein, the inhibition of host macromolecular synthesis appears to be a prerequisite for the synthesis of structural components of the virus (Aurelian and Roizman, 1965). It could be that tumor-associated herpesviruses have the same requirement but are unable to terminate unassisted the biosynthesis of host macromolecules. However, quite clearly the mechanism postulated for the control of viral macromolecular synthesis in cells harboring herpes simplex virus is not applicable to cells harboring the tumor-associated viruses.

Are the tumor-associated herpesviruses responsible for the tumors in which they are found? The answer is not clear, and evidence other than that which has been cited is very meager. However, as pointed out earlier, the virus associated with Burkitt lymphoma or viruses very closely related to it are widespread and readily isolated throughout the world. Yet the lymphoma described by Burkitt and postnasal carcinoma are uncommon outside a circumscribed area of Africa and outside people of Chinese extraction, respectively. We cannot exclude the possibility that the specific cell harboring the virus and the expression of some viral genes—sufficient to transform the cell—are singly or both determined by the genotype of their host.

Some weeks ago there appeared in the U.S. press a statement by a noted pediatrician to the effect that too much salt and sugar can be very dangerous for infants. The statement was brief. It was not clear whether salt and sugar were the only items or whether somehow only the items arranged alphabetically under the letter S made the news. Since the article appeared shortly after baby food manufacturers had decided to omit monosodium glutamate from infant food, I rather suspect these were two items on a very long list. If this portends the diet of the future, I am glad my baby food days are over. This, however, is beside the point. This essay half-frivolously referred to herpesviruses as viruses of love. The half that is not frivolous is based on the accumulated evidence that herpesviruses as a group are transmitted most readily by personal contacts and particularly by kissing and sexual intercourse. It is perhaps fitting that in closing I should paraphrase the pediatrician thus: too much love can also be dangerous. Alas!

REFERENCES

Antonelli, A., and C. Vignali (1968). Ricerche sulla localizzazione del virus dell' herpes simplex dopo guanigione dele recidive cutanee. Coltivazione di cellule della zona colpites da enzione recidivante, *Riv. Ist Sieroter Italiano,* **43,** 43–51.

Ashe, W. K., and H. W. Scherp (1965). Antigenic variations in herpes simplex virus isolants from successive recurrences of herpes labialis, *J. Immunol.*, 94, 385–394.

Aurelian, L. (1968). Effect of environmental conditions on formation and release of canine herpesvirus in infected canine kidney cells, *Amer. J. Vet. Res.*, 29, 1945–1952.

Aurelian, L., and B. Roizman (1965). Abortive infection of canine cells by herpes simplex virus. II. The alternative suppression of synthesis of interferon and viral constituents, *J. Mol. Biol.*, 11, 539–548.

Bender, C. E. (1962). Recurrent mononucleosis *J.A.M.A.*, 182, 954–956.

Ben-Porat, T., and A. S. Kaplan (1965). Mechanism of inhibition of cellular DNA synthesis by pseudorabies virus, *Virology*, 25, 22–29.

Blank, H., and M. W. Brody (1950). Recurrent herpes simplex: a psychiatric and laboratory study, *Psychosomat. Med.*, 12, 254–260.

Blank, H., and G. W. Rake (1965). *Viral and Rickettsial Diseases of the Skin, Eye and Mucous Membranes of Man*, Boston, Little, Brown and Co.

Bowling, C. P., C. R. Goodheart, and G. Plummer (1969). Oral and genital bovine herpesviruses, *Virology*, 3, 95–97.

Buddingh, J., D. I. Schrum, J. C. Lanier, and D. J. Guidry (1953). Studies on the natural history of herpes simplex infections, *Pediatrics*, 11, 595–610.

Burnet, F. M., and S. W. Williams (1939). Herpes simplex: a new point of view. *Med. J. Australia*, 1, 637–642.

Carton, C. A. (1953). Effect of previous sensory loss on the appearance of herpes simplex following trigeminal sensory root section, *J. Neurosurgery*, 10, 463–468.

Carton, C. A., and E. D. Kilbourne (1952). Activation of latent herpes simplex by trigeminal sensory-root section, *New Eng. J. Med.*, 246, 172–176.

Causey, G. (1960). *The Cell of Schwann*, Livingstone, Edinburgh, pp. 86–94.

Coriell, L. L. (1963). Discussion of the paper by B. Roizman, in *Virus, Nucleic Acids and Cancer*, Baltimore, Williams & Wilkins Co., p. 241.

Cushing, H. 1905. The surgical aspects of major neuralgia of the trigeminal nerve. A report of twenty cases of operation on the gasserian ganglion with anatomic and physiologic consequences of its removal, *J.A.M.A.*, 44, 773–779, 800–805, 920–929, 1002–1008, 1088–1093.

Darlington, R. W., and L. H. Moss (1968). III. Herpesvirus envelopment. *J. Virol.*, 2, 48–55, 1968.

Doerfler, W. (1968). The fate of the DNA of adenovirus type 12 in baby hamster kidney cells, *Proc. Nat. Acad. Sci. U.S.*, 60, 636–643.

Doerr, R. (1938). Herpes febrilis, in *Handbuch der Virusforschung*, vol. 1, pp. 41–45, Vienna, Springer-Verlag.

Dowdle, W. R., A. J. Nahmias, R. W. Harwell, and F. P. Pauls (1967). Association of antigenic type of herpesvirus hominis with site of viral recovery, *J. Immunol.*, 99, 974–980.

Ejercito, P. M., E. D. Kieff, and B. Roizman (1968). Characterization of herpes simplex virus strains differing in their effect on social behavior of infected cells, *J. Gen. Virol.*, 3, 357–364.

Ellison, S. A., C. A. Carton, and H. M. Rose (1959). Studies of recurrent herpes simplex infections following section of the trigeminal nerve, *J. Infect. Dis.*, 105, 161–167.

Epstein, M. A., B. G. Achong, and Y. M. Barr (1964). Virus particles in cultured lymphoblasts from Burkitt's lymphoma, *Lancet,* 2, 702–703.

Evans, A. S. (1960). Infectious mononucleosis in University of Wisconsin students. Report of a five-year investigation, *Amer. J. Hyg.,* 71, 342–362.

Evans, A. S. (1963). Recurrent mononucleosis, *J.A.M.A.,* 184, 515.

Falchi, G. (1925). Herpes sperimentale recidivante nell'uomo, *Boll. Soc. Medico-Chirurg. Pavia,* 37, 885.

Fawcett, O. W. (1956). Electron microscopic observations on intracellular virus-like particles associated with the cells of the Lucké renal adenocarcinoma, *J. Biophys. Biochem. Cytol.,* 2, 725–742.

Feldman, L. A., R. D. Sheppard, and M. B. Bornstein (1968). Herpes simplex virus-host cell relationships in organized cultures of mammalian nerve tissues, *J. Virol.,* 2, 621–628.

Figueroa, M. E., and W. E. Rawls (1969). Biological markers for differentiation of herpes virus strains of oral and genital origin, *J. Gen. Virol.,* 4, 259–267.

Findloy, G. M., and F. O. MacCallum (1940). Recurrent traumatic herpes, *Lancet,* 238, 259–261.

Goodheart, C. R., G. Plummer, and J. L. Waner (1968). Density difference of DNA of human herpes simplex viruses types I and II, *Virology,* 35, 473–475.

Goodpasture, E. W. (1929). Herpetic infection, with especial reference to involvement of the nervous system, *Medicine,* 8, 223–243.

Hampar, B., and M. A. K. Burroughs (1969). Mechanism of persistent herpes simplex virus infection *in vitro, J. Nat. Cancer Inst.,* 43, 621–634.

Heilig, R., and H. Hoff (1928). Ueber psychogenen eustehuug des herpes labialis, *Med. Klin.,* 24, 1472.

Henle, W., and G. Henle (1968). Effect of arginine-deficient media on the herpes-type virus associated with cultured Burkitt tumor cells, *J. Virol.,* 2, 182–191.

Henle, G., W. Henle, and V. Diehl (1968). Relation of Burkitt's tumor-associated herpes-type virus to infectious mononucleosis, *Proc. Nat. Acad. Sci. U.S.,* 59, 94–101.

Hoagland, R. J. (1952). The transmission of infectious mononucleosis, *Amer. J. Med. Sci.,* 229, 262–272.

Hsiung, G. D., and L. S. Kaplow (1969). Herpeslike virus isolated from spontaneously degenerated tissue culture derived from leukemia-susceptible guinea pig, *J. Virol.,* 3, 355–357.

Josey, W. E., A. J. Nahmias, and Z. M. Naib (1968). Genital infection with type 2 herpesvirus hominis—present knowledge and possible relation to cervical cancer, *Amer. J. Obst. Gyn.,* 101, 718–729.

Kaufmann, H. E., D. C. Brown, and E. E. Maloney (1967). Recurrent herpes in the rabbit and man, *Science,* 156, 1928.

Keller, J. M., P. G. Spear, and B. Roizman (1970). The proteins specified by herpes simplex virus. III, in press.

Kvasnicka, A. (1963). Relationship between herpes simplex and lip carcinoma. II. Antiherpetic antibodies in patients with lip cancer, *Neoplasma (Bratislava),* 10, 82–88.

Kvasnicka, A. (1964). Relationship between herpes simplex and lip carcinoma. III. *Neoplasma (Bratislava),* 10, 199–203.

Kvasnicka, A. (1965). Relationship between herpes simplex and lip carcinoma. IV. Selected cases, *Neoplasma (Bratislava),* 12, 61–70.

Leider, W., R. L. Hagoffin, E. H. Lenette, and L. N. R. Leonards (1965). Herpes simplex virus encephalitis, *New Eng. J. Med.,* **273,** 341–347.

Love, R., A. S. Rabson, and P. Wildy (1964). Changes in the nucleolus of normal and neoplastic cells infected with riboviruses and deoxyriboviruses, *Acta Unio Intern. Contra Cancrum,* **20,** 1384–1387.

Lucké, B. (1938). Carcinoma of the leopard frog: its probable causation by a virus, *J. Exp. Med.,* **68,** 457–466.

Lunger, P. D. (1964). The isolation and morphology of the Lucké frog kidney tumor virus, *Virology,* **24,** 138–145.

Luse, S., P. Friedman, and M. Smith (1965). An ultrastructural study of herpes simplex encephalitis, *Amer. J. Path.,* **46,** 8A.

Lwoff, A. (1953). Lysogeny, *Bact. Rev.,* **17,** 269–337.

Melendez, L. V., R. D. Hunt, M. D. Daniel, F. G. Garcia, and C. E. O. Fraser (1969). Herpesvirus Saimiri. II. Experimentally induced malignant lymphoma in primates, *Lab. Animal Care,* **19,** 378–386.

Mizell, M. (1969). Introduction: State of the art. *Proceedings of the Symposium on the Biology of Amphibian Tumors,* Berlin, Springer-Verlag.

Morgan, J. K. (1968). Herpes gestationis influenced by an oral contraceptive. *Brit. J. Derm.,* **80,** 456–458.

Nahmias, A. J., W. Chiang, I. del Buono, and C. Duffey (1969). Typing of Herpes hominis strains by immunofluorescent techniques, *Proc. Soc. Exper. Biol. Med.,* **132,** 386.

Nahmias, A. J., and W. R. Dowdle (1968). Antigenic and biologic differences in herpesvirus hominis, *Progr. Med. Virol.,* **10**

Nahmias, A. J., S. Kibrick, and R. C. Rosan (1964). Viral leukocyte inter-relationships. I. Multiplication of a DNA virus—herpes simplex—in human leukocyte cultures, *J. Immunol.,* **93,** 69–74.

Nazerian, K., J. J. Solomon, R. L. Witter, and B. R. Burmester (1968). Studies on the etiology of Marek's disease. II. Finding of a herpesvirus in cell culture, *Proc. Soc. Exp. Biol. Med.,* **127,** 177–182.

Nicolau, S., and P. Poincloux (1928). Resultats d'une greffe de plaue saine faite en 1924 sur un doigt atteint d'herpes recidivant pendant les cinq années précédentes, *Compt. rend. Soc. Biol. (Paris),* **98,** 360.

Paine, T. F. (1964). Latent herpes simplex infection in man, *Bact. Rev.,* **28,** 472–479.

Patrizi, G., J. N. Middelkamp, and C. A. Reed (1968). Fine structure of herpes simplex virus hepatoadrenal necrosis in newborn, *Amer. J. Clin. Path.,* **49,** 325–341.

Plummer, G., J. L. Waner, and C. P. Bowling (1968). Comparative studies of type 1 and type 2 "herpes simplex" viruses, *Brit. J. Exp. Path.,* **49,** 202–208.

Rabin, E. R., A. B. Jenson, C. A. Phillips, and J. L. Melnick (1968). Herpes simplex virus hepatitis in mice: an electron microscopic study, *Exp. Mol. Path.,* **8,** 34–48.

Rafferty, K. A., Jr. (1964). Kidney tumors of the leopard frog: a review. *Cancer Res.,* **24,** 169–185.

Rawls, W. E., W. A. Tompkins, M. E. Figueroa, and J. L. Melnick (1968). Herpesvirus type 2 association with carcinoma of cervix, *Science,* **161,** 1255–1266.

Roane, P. R., Jr., and B. Roizman (1964). Studies of the determinant antigens of viable cells. II. Demonstration of altered antigenic reactivity of HEp-2 cells infected with herpes simplex virus, *Virology,* 22, 1–8.

Roizman, B. (1962). Polykaryocytosis, *Cold Spring Harbor Symp. Quant. Biol.,* 27, 327–342.

Roizman, B. (1963). The programming of herpes virus multiplication in mammalian cells, in *Viruses, Nucleic Acids, and Cancer* (Proceedings of the 17th Annual Symposium. M. D. Anderson Hospital and Tumor Institute), Baltimore, Williams & Wilkins Co., pp. 205–223.

Roizman, B. (1965). An inquiry into the mechanisms of recurrent herpes infections of man, in *Perspectives in Virology,* vol. IV, pp. 283–304, edited by M. Pollard, New York, Harper & Row, Hoeber Medical Division.

Roizman, B. (1969a). The herpesviruses—a biochemical definition of the group, in *Current Topics in Microbiology and Immunology,* vol. 49, pp. 1–79, Heidelberg, Springer-Verlag.

Roizman, B. (1969b). Herpesviruses, membranes, and the social behavior of infected cells, in *Proceedings of the International Symposium on Medicine and Applied Virology,* in press.

Roizman, B., and L. Aurelian (1965). Abortive infection of canine cells by herpes simplex virus. I. Characterization of viral progeny from cooperative infection with mutants differing in ability to multiply in canine cells, *J. Mol. Biol.,* 11, 528–538.

Roizman, B., and P. R. Roane, Jr. (1963). Demonstration of a surface difference between virions of two strains of herpes simplex virus, *Virology,* 19, 198–204.

Roizman, B., and P. R. Roane, Jr. (1964). The multiplication of herpes simplex virus. II. The relation between protein synthesis and the duplication of viral DNA in infected HEp-2 cells, *Virology,* 22, 262–269.

Roizman, B., and P. G. Spear (1968). Preparation of herpes simplex virus of high titer. *J. Virol.,* 2, 83–84.

Roizman, B., and S. B. Spring (1967). Alteration in immunologic specificity of cells infected with cytolytic viruses, in *Proceedings of the Conference on Cross-Reacting Antigens and Neoantigens,* edited by J. J. Trentin, Baltimore, Williams & Wilkins Co., pp. 85–96.

Roizman, B., S. B. Spring, and J. Schwartz (1969). The herpesvirion and its precursor made in productively and abortively infected cells (Proceedings of the Symposium on Viral Defectiveness), *Fed. Proc.,* November–December.

Rustigian, R., J. B. Smulow, M. Tye, W. A. Gibson, and E. Shindell (1966). Studies on latent infection of skin and oral mucosa in individuals with recurrent herpes simplex, *J. Inves. Derm.,* 47, 218–221.

St. Gene, J. W., J. T. Prince, B. A. Burke, R. A. Good, and W. Krivit (1965). Impaired cellular resistance to herpes-simplex virus in Wiskott-Aldrich syndrome, *New Eng. J. Med.,* 273, 229–234.

St. Gene, J. W., Jr., J. T. Prince, B. A. Burke, and W. Krivit (1962). Studies of persistent herpes virus infection in children with Aldrich syndrome, *J. Pediat.,* 61, 302.

Schneck, J. M. (1944). The psychological components in a case of herpes simplex, *Psychosomat. Med.,* 3, 349.

Schneweiss, K. E. (1962). Serologische Untersuchungen zur Typedifferzierung des Herpesvirus hominis, *Z. Immunoforsch.,* 124, 24–48.

Schwartz, J., and B. Roizman (1969a). Concerning the egress of herpes simplex virus from infected cells: electron microscope observations, *Virology*, 38, 42–49.

Schwartz, J., and B. Roizman (1969b). Similarities and differences in the development of laboratory and of freshly isolated herpes simplex virus strains in HEp-2 cells. Electron microscopic studies, *J. Virol.*, December.

Scott, T. F. McN. (1957). Epidemiology of herpetic infections, *Amer. J. Ophthal.*, 43, 134–147.

Selling, B., and S. Kibrick (1964). An outbreak of herpes simplex among wrestlers (herpes gladiatorum). *New Eng. J. Med.*, 270, 979–982.

Spear, P. G., J. M. Keller, and B. Roizman (1970). The proteins specified by herpes simplex virus. II. Viral glycoproteins associated with cellular membranes, *J. Virol.*, in press.

Spear, P. G., and B. Roizman (1968). The proteins specified by herpes simplex virus. I. Time of synthesis, transfer into nuclei, and properties of proteins made in productively infected cells, *Virology*, 36, 545–555.

Stalder, W., and S. Zurukzoglu (1936). Experimentelle Untersuchungen über Herpes. Transplantation herpesinfizierter Hautstellen, Reaktivierung von abgeheilten, künstlich infizierten Hautstellen. Herpesbehandlung. VI, *Mitt. Zentralbl. Bakt.* (I Abt. Orig.), 136, 94–97.

Swanson, J. L., J. E. Craighead, and E. S. Reynolds (1966). Electron microscopic observations on herpesvirus hominis (herpes simplex virus) encephalitis in man, *Lab. Invest.*, 15, 1966–1981.

Sydiskis, R. J., and B. Roizman (1967). The disaggregation of host polyribosomes in productive and abortive infection with herpes simplex virus, *Virology*, 32, 678–686.

Tankersley, R. W., Jr. (1964). Amino acid requirements of herpes simplex virus in human cells, *J. Bact.*, 87, 609–613.

Terni, M. (1965). L'Infezione erpetica recidivante: conoscenze E problemi, *L'Arcispedale S. Anna di Ferrara*, 18, 515–532.

Terni, M., and B. Roizman (1970). Variability of herpes simplex virus: isolation of two variants from simultaneous eruptions at different sites, *J. Infect. Dis.*, February.

Tokumaru, T. (1966). A possible role of A-immunoglobulin in herpes simplex virus infection in man, *J. Immunol.*, 97, 248–259.

Tweedell, K. S. (1967). Induced oncogenesis in developing frog kidney cells. *Cancer Res.*, 27, 2042–2052.

Wagner, E. K., and B. Roizman (1969). RNA synthesis in cells infected with herpes simplex virus. I. The patterns of RNA synthesis in productively infected cells, *J. Virol.*, 4, 36–46.

Warren, S. L., C. M. Carpenter, and R. A. Boak (1940). Symptomatic herpes, a sequela of artificially induced fever; incidence and clinical aspects; recovery of virus from herpetic vesicles, and comparison with known strain of herpes virus, *J. Exp. Med.*, 71, 155–168.

Wheeler, G. E., Jr., and W. H. Cabaniss (1965). Epidemic cutaneous herpes simplex in wrestlers (herpes gladiatorum), *J.A.M.A.*, 194, 993–997.

Wight, P. A. L., J. E. Wilson, J. G. Campbell, and E. Fraser (1967). Virus-like particles in blood lymphocytes in acute Marek's disease, *Nature*, 216, 804–805.

Wyburn-Mason, R. (1957). Malignant change following herpes simplex. *Brit. Med. J.*, 2, 615–616.

The gene, tox⁺, of Corynebacterium diphtheriae

LANE BARKSDALE

The gene *tox⁺* (3) of *Corynebacterium diphtheriae* governs the synthesis of diphtherial toxin, a protein[1] of molecular weight about 64,000 (29, 31, 44, 46, 47, 49, 51), lethal for man and animals in doses of 130 ng/kg body weight (5, 60). Although *tox⁺* must have plagued man since ancient times, it was not until 1826, when Pierre Brettoneau described the clinical entity, diphtheria, that meaningful recording of this disease began. After the etiologic agent, *Corynebacterium diphtheriae,* had been discovered by Klebs in 1883 (32) and related to the disease by Loeffler in 1884 (34), a rational means was available for distinguishing diphtheria from other maladies of the throat. In 1888, Roux and Yersin (54) made the exciting discovery that toxic filtrates could be easily obtained free of diphtheria bacilli and that these filtrates were lethal for animals. Since the symptoms produced by toxin in animals (neurologic changes and cardiac failure) accounted for the more dramatic signs seen in human diphtheritic infections, the action of toxin was held to account for the pathogenesis of diphtheria, and diphtheria bacilli which lacked *tox⁺* were all but ignored and their relationship with *C. diphtheriae* was little considered.

In 1896 William Hallock Park isolated a mutant of the diphtheria

[1] Diphtherial toxin was obtained in crystalline form (needles, plates, or shields) by Pope and Stevens in 1953 and by Katsura, Kato, Nakamura, and Koyama in 1957.

bacillus, the PW8 strain,[2] which was a hyperproducer of toxin (45). By 1925, after the demonstration by Ramon (48) that formalin-treated toxin (*anatoxin, toxoid*) was a nontoxic agent suitable for rendering animals immune to diphtherial toxin, the PW8 strain came to be used for the production of toxin in many laboratories around the world. From it we have learned a number of facts about the *tox*+ gene.

When Freeman discovered in 1951 that certain bacteriophages could endow nontoxinogenic *C. diphtheriae* with the capacity to produce toxin (16), interest in the biology of *C. diphtheriae* was rekindled. Soon one of the temperate phages (7, 20, 21) carrying *tox*+, β^{tox+}, was characterized, and pairs of toxinogenic and nontoxinogenic strains of *C. diphtheriae*, such as $C7_s(-)^{tox-}$, $C7_s(\beta)^{tox+}$, $C4_s(-)^{tox-}$, and $C4_s(\beta)^{tox+}$, were isolated, cloned, and studied (4, 20). For the first time in the 68 years since their discovery, diphtheria bacilli differing by only one gene could be compared. It was at once obvious that invasiveness (virulence) and toxinogenicity were separable properties (58). For example, rabbits infected with $C7_s(-)^{tox-}$ developed pseudomembranous lesions but later recovered from their infections, whereas rabbits infected with $C7_s(\beta)^{tox+}$ developed necrotic lesions and died (5).

This separation of invasiveness from toxinogenicity was consistent with the impressions many people had had regarding the possibility of diphtheritic infections in man caused by nontoxinogenic diphtheria bacilli and infections in individuals having circulating antitoxin (8, 14, 27). Once it was clear that the already-known wide variety of antigenic types of *C. diphtheriae* (6, 14, 52), toxinogenic or not, had the capacity to colonize man, it was obvious that they also could be responsible for the "nonspecific" reactions to the Schick test antigen(s) seen, especially in nonimmunized individuals (8, 27).[3] The toxin and toxoid used in the

---

[2] The PW8 strain carries a prophage P and correctly is designated $PW8_r(P)$ or $PW8_r(Pd)$. The prophage was originally designated Pd because, after induction with ultraviolet light, lysis occurred but only a very occasional plaque-forming particle was released. Furthermore, the amount of phage DNA synthesized in the induced cell (compare Figs. 5 and 6) was small compared to normally lysogenic corynebacteria. This was further complicated by the finding that such P-phage is restricted in the standard indicator strain, $C7_s(-)^{tox-}$, but not in *C. ulcerans* 603 (Ref. 33; see also Ref. 39). Even so, in any induced population of $PW8_r(Pd)$ less than 10% of the cells yield phage.

[3] The immune status of an individual, with respect to diphtherial toxin, may be determined through the use of the modified Schick test. Approximately 0.0008 μg

Schick test are commonly prepared from the Park Williams Number 8 strain. Although the toxin is usually purified by precipitation with ammonium sulfate, it does contain a number of antigens of the PW8 bacillus which are common to several other strains of *C. diphtheriae* (50). Therefore, persons who have had an immune response to any of these cross-reacting strains can be expected to give one of the several reactions possible in the Schick test, including the so-called pseudoreaction.

CORYNEBACTERIOPHAGES AND THE LOCUS FOR *tox*$^+$

Holmes (24) has recently worked out a mating system for corynebacteriophages and, using the markers *h* (host range), *imm* (lysogenic immunity), *tox*$^+$, *c* (clear plaque), and *h'* (extended host range), has shown that, in crosses between phages β^{tox+} and γ^{tox-}, *tox*$^+$ behaves as though it were close to *h* and a map order of—*h*—*tox*$^+$—*imm*β—*c*—*h'*—seems well established for phage β^{tox+} (see also Fig. 1).

More recently Holmes (25) has examined morphologically and serologically distinct phages carrying *tox*$^+$ for their capacity to undergo genetic recombination as a measure of their relatedness. Although it was expected that all *tox*$^+$ phages would probably undergo genetic recombination, this was not the case. In fact, *tox*$^+$ is found in bacteriophages which are morphologically distinct, in phages which are serologically distinct, and among phages which cannot recombine genetically. When hybrid phages differing only in the presence or the absence of *tox*$^+$ are compared as to efficiency of absorption, latent period, burst size, stability to storage, etc., they seem to be alike. In other words, *tox*$^+$ seems to endow phages with no special advantage(s).

---

of toxin protein, or $\frac{1}{50}$ of the minimal amount required to kill a guinea pig weighing 250 g, is injected into the skin of the forearm. An equal amount of toxoid is injected at a control site. Necrosis at the test site indicates a nonimmune state, while immunity, the presence of circulating antitoxin, is indicated by a lack of a reaction at either site. Immunity complicated by allergy to corynebacterial proteins results in a delayed (but sometimes immediate) inflammatory reaction at both sites.

Figure 1. This rendering of corynebacteriophages and the map position of *tox*+, by James E. Ziegler (in the form of a printer's end piece), is in the spirit of the time-span of Lwoff-in-biology: 50 years which bridge the era of observe-draw-and-stipple and the era of get-out-the-relevant-molecule.

STABILITY OF THE *tox*⁺ PROPHAGES

Although it has been reported that toxinogenic strains of *C. diphtheriae* may be rendered nontoxinogenic by loss of their *tox*+-containing prophages (1), our experience has been that such prophages are very stably integrated and that only in pseudolysogenic strains (35), carrier cultures, is loss of toxinogenicity observed. In fact, no prophage loss has been observed in strains of $C4_s(\beta)^{tox+}$ and $C7_s(\beta)^{tox+}$ which have been under cultivation in our laboratory for over 18 years. In this connection the PW8 strain is especially interesting. It is a rough strain lacking receptors for any of the known corynebacteriophages. Therefore, should it lose its *tox+*-carrying prophage, reinfection is unlikely. This bacterium has been in continuous cultivation for 74 years. Recently, we examined five strains of PW8 maintained in laboratories in various parts of the world and found them all to be toxinogenic and lysogenic (33). Thus the corynebacterioprophages examined seem stably integrated with the host genome and offer some of the best examples of this kind of stability on record.

EXPRESSION OF *tox+*

Diphtherial toxin can be produced by nontoxinogenic diphtheria bacilli infected with virulent *tox+* phages, as well as by lysogenic, toxinogenic *C. diphtheriae*. There are available, then, two systems for studying the expression of *tox+*.

Any discussion of the biosynthesis of toxin must begin with a consideration of iron, the *bête rouge* of the *tox+* story (15, 42, 43, 61). First, certain *beliefs* about iron and toxin production need to be reviewed. The impression has been (*a*) that iron blocks toxin production and (*b*) that toxin is made only when diphtheria bacilli "run out of iron" (4). However, iron does not completely block toxin production; small amounts of toxin are made in the presence of excess iron (42). Lysogenic, toxinogenic diphtheria bacilli, which are making toxin, have not run out of iron. They are existing in a narrow range of iron which permits active metabolism but to an increasing degree (with time) restricts the process of cell division.

FERRUMINATION AND FERROTYPES

Figure 2 presents growth curves of populations of diphtheria bacilli initially grown in levels of iron suitable for division at the maximal rate and then washed and subcultured (*a*) in the presence of 0.15 μg/ml added iron and (*b*) in the absence of added iron. While both groups of cells exhibit a "log phase" or growth, only the cells which are grow-

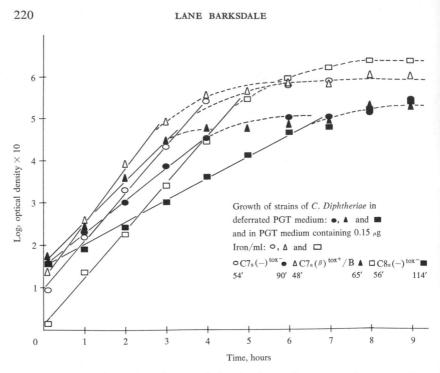

Figure 2. The effect of the iron available in the medium upon the generation time of various strains of *Corynebacterium diphtheriae*. (From unpublished data of Ernest Bell and L. Barksdale.) See also Figs. 3 and 4.

ing in 0.15 µg/ml iron are exhibiting the maximal rate of division established for these three strains of *C. diphtheriae*. If one grows three populations of diphtheria bacilli in the presence of (*a*) 1.0 µg added iron/ml, (*b*) 0.075 µg added iron/ml, and (*c*) 0.0 µg added iron/ml and then twice washes them in deferrated medium, resuspends them in deferrated medium, and follows their growth, the amount of growth observed will reflect the *amount of iron reserves bound by the cells of each population*. In Fig. 3 it is apparent that even the most iron-starved population is capable of an increase in optical density *equivalent to* two divisions, albeit at a rate much slower than normal. When these cells are examined under the light microscope, they are seen to be much longer than normal cells. They are not dividing. This *failure in division* is the *hallmark* of the *induced lysogenic cell* and of the *iron-starved cell*.

Righelato and van Hemert (53) have recently examined the syn-

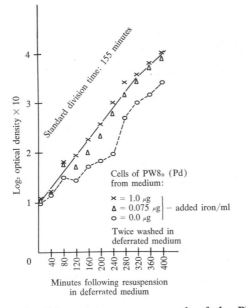

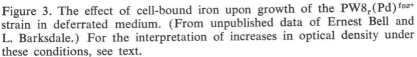

Figure 3. The effect of cell-bound iron upon growth of the PW8$_r$(Pd)$^{tox+}$ strain in deferrated medium. (From unpublished data of Ernest Bell and L. Barksdale.) For the interpretation of increases in optical density under these conditions, see text.

thesis of toxin by batch- and chemostat-grown cultures of the PW8 strain CN2000, and Righelato (52) has looked into the matter of the distribution of iron in cells growing with small amounts of iron (7 μg-atom/l) and excess amounts (97 μg-atom/l). In the low-iron cells (LFC) the amount of iron found in heme-containing enzymes was only 13% of the iron bound by the cells. The high-iron cells (HFC), on the other hand, took up 15 times more iron (but only 80% of the total amount available to them), 3.7% of which was heme iron. Here, then, are two different phenotypes of the diphtheria bacillus, marked by different ratios of heme to nonheme iron. One cell makes large amounts of toxin; the other, very little. It is interesting that there are ultrastructural differences between these two phenotypes. If one examines ultrathin sections of the low-iron and high-iron cells (see Fig. 4), these differences are apparent. The HFC have more layers in the cell envelope and contain numerous electron-opaque areas. The LFC have fewer or no electron-opaque areas and exhibit a much simpler cell envelope,

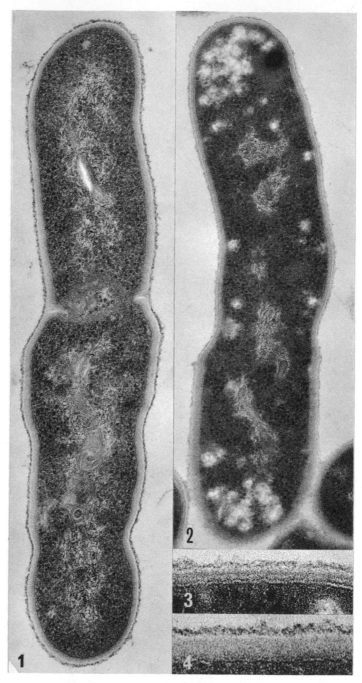

Figure 4

which we are inclined to think is related in some way to the release of toxin.

FERROTYPES AND YIELDS OF TOXIN

There are two categories of toxin production, and these are related to the iron phenotypes, LFC and HFC: (*a*) the HFC produce small yields (3 mg/ml or less), termed zero level by some investigators (52, 53), and (*b*) the LFC produce large yields (10–300 mg/ml). Though many workers have ignored category (*a*), both categories are important to understanding the expression of *tox*+.

The most informative experiments concerning these phenotypes and toxin production are those in which Matsuda (37, 38) infected LFC and HFC of nontoxinogenic $C7_s(-)^{tox-}$ with phage β^{hvtox+} and followed the appearance of intra- and extracellular toxin. He found the rate of synthesis of toxin to be essentially the same for the two infected phenotypes, but the final accumulation of toxin in the extracellular milieu was greater in the LFC population. It seems, therefore, that release of toxin is curtailed in HFC cells.

Whereas in the Matsuda experiments toxin is being synthesized in conjunction with bacteriophage growth, the amount of the contribution of vegetative phage growth to the expression of *tox*+ in lysogenic, toxinogenic cultures depends on whether the population is HFC or LFC. In our experience with the HFC of a variety of lysogenic, toxinogenic *C. diphtheriae,* toxin is always found in a range of 200–2000 molecules/cell. This is from $\frac{1}{50}$ to $\frac{1}{5}$ the maximal amounts found in LFC populations. Since cells in which *tox*+ phages multiply make toxin at a maximal rate (37, 38) and since the accumulated lyses in any population of toxinogenic diphtheria bacilli are proportional to growth, the toxin produced in high-iron cultures is, almost certainly, that made in cells in which phage biosynthesis is taking place. Furthermore, when cells in the transition from HFC to LFC are treated with inducing doses of ultraviolet light, they begin to make toxin long (5 hours) before the untreated control cultures (4).

←Figure 4. (1) Low-iron cell (LFC) and (2) high-iron cell (HFC) of $Pw8_r(Pd)^{tox+} \times 60,000$, (3) segment of cell envelope of HFC, and (4) of $LFC \times 178,000$. Cells were fixed in glutaraldehyde and osmium and embedded in epon. (From unpublished data of Kwang Shin Kim and L. Barksdale.)

As the control cultures begin to make toxin, they take on the appearance of LFC and their extracellular toxin levels increase linearly with time. In 1960, Yoneda and Barksdale (62) attempted to see whether or not induction of prophage P in LFC populations of the PW8 strain would lead to a boost in the rate of synthesis of toxin. This bacterium is exquisitely sensitive to mitomycin C, and exposure to 0.1 μg/4 $\times$ 10^8 bacteria for 60 minutes yields optimal induction of its carried prophage. It was found that after the induction of prophage P there was, as expected, an arrest of bacterial DNA synthesis and a slight rise of the induced P-phage DNA, but this was not followed by any change in the rate of toxin synthesis. The experimenters were struck by the fact that essentially moribund cells went on synthesizing toxin.

Hirai and his colleagues thereafter examined dividing and non-dividing populations of the PW8 strain and found both capable of producing toxin (23). Later, Matsuda and Barksdale compared the production of toxin by several strains of *C. diphtheriae* induced with mitomycin C. Results from two of their experiments are shown in Figs. 5 and 6. It is clear from these data that in cells which are incapable of further division and in which bacterial DNA synthesis has been blocked RNA synthesis and protein synthesis continue at one-half the rate of the control cells. Toxin, then, can be synthesized by cells which are no longer capable of vegetative reproduction. This leaves us with the inescapable fact that in the LFC population *tox+* is turned on and synthesis of toxin continues as long as there is RNA synthesis. Whether all of the cells or only a fraction are responsible for the toxin synthesized remains a question.

tox+: STRUCTURAL GENE OR NOT?

If *tox+* is a structural gene and toxin the product of its message, then the synthesis of toxin over a period of 10 hours in the absence of appreciable DNA synthesis (see Fig. 5) suggests that the *tox+* messenger is indeed stable. On the other hand, if the product of *tox+* expression is an enzyme with a reasonable turnover capacity, the continued synthesis of toxin (see Figs. 5 and 6) would then be secondary to messenger synthesis. Until one knows what toxin is, it remains difficult to assess the role of *tox+*.

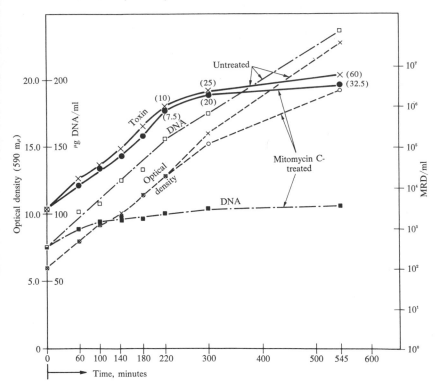

Figure 5. The increase in optical density and the synthesis of toxin by cultures of the PW8ᵣ(Pd)*ᵗᵒˣ⁺* strain of *C. diphtheriae* treated and not treated with mitomycin C. Exposure to mitomycin C, 0.1 μg/O.D. 0.3, was for 1 hour. Cells were then washed and resuspended in fresh medium. Standard procedures were employed for the determination of DNA, RNA, protein and toxin; for details see Ref. 38. To simplify this figure, data on RNA and protein have been omitted. The values obtained were essentially those shown in Fig. 6. Toxin was measured as MRD when the levels were small, and as flocculating units, L_f, when the levels were above 8 μg toxin protein/ml. The L_f units are shown in parentheses: 10^5 MRD ≈ 1 L_f ≈ 1.75 μg toxin protein. (These data are from unpublished work of Matsuda and Barksdale. For media and other details, see Ref. 38.)

Although there is as yet no understanding of what toxin is, there is considerable information as to what it is not. For example, it has been proposed that *tox+* was linked to the nitratase marker of *C. diphtheriae* and that when phage β^{tox+} was integrated into the genomes of such nitratase-negative (*Nred-*) corynebacteria as *C. ulcerans* and *C.*

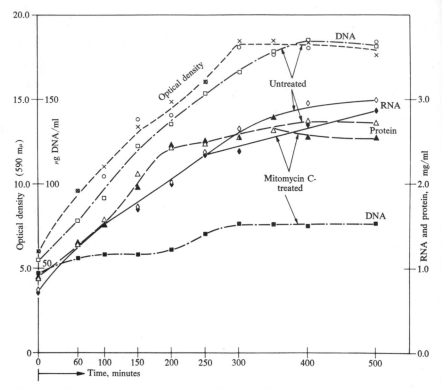

Figure 6. Increase in optical density, DNA, RNA, and protein in mitomycin C-treated and untreated cultures of $PW8_s(P)^{tox+}$. The maximal level of toxin produced in this experiment was approximately 11 μg/ml. For other details see Fig. 5 and the text.

belfanti (21, 22), these bacteria became nitratase positive and *tox*+. While these experiments do not seem to be reproducible with the strains originally employed, Arden and co-workers (2, 17) have shown that, when the phage β^{tox+}, l^{tox+}, or β^{hvtox+} is introduced into nitratase-negative strains such as *C. ulcerans* 603 and $C7_s^{Nred-}$, the resultant lysogenic strains and/or lysates are toxin producers and are nitratase negative. The interesting observation by Warren and Spearing (59) of an association between diphtherial toxin and neuraminidase activity, and antitoxin and antineuraminidase activity, posed the question of whether or not toxin might originate from the formation of dimers or trimers of neuraminidase lacking the capacity to turn over their substrates. Moriyama (41) has done a careful study of this possibility and has found no link between neuraminidase and diphtherial toxin.

The studies on the effect(s) of toxin on HeLa cells, begun by Strauss and Hendee (56, 57) and continued and expanded by Pappenheimer and his collaborators, have given the first inkling of what "purified preparations" of toxin may do to animal cells. Strauss and Hendee established that (a) in intoxicated HeLa cells glycolysis and aerobic respiration continued at a normal rate for many hours, whereas (b) protein synthesis, as reflected in the inability to incorporate radioactive methionine, was very early stopped; also, (c) the intoxicated cells developed visible blebs within 4 hours and underwent destruction about 7 hours after exposure to toxin. Pappenheimer and his collaborators (12, 19, 30) have accumulated evidence to show that "purified toxin" can tie up protein synthesis in in vitro mammalian systems, and Collier (10) and Honjo et al. (26) have shown that toxin does this by bringing about the ribosylation of transferase 2 with the ADPR moiety of NAD.

Goor (18) and Collier (11) have independently shown that more than one form of toxin exists. Collier's findings are in keeping with the observation of Bizzini et al. (9) that the sedimentation coefficient of toxin is reduced from 4.2S to 2.0S after treatment with disulfite and the finding of Iskierko (28) that toxin appears to have more than one equivalent of amino-terminal acid per 65,000 molecular weight. Collier (13) has shown that it is the subunits and not toxin which bring about the hang-up in the protein synthesizing systems. The subunits, 2.5S as opposed to 4.2S for toxin, (a) are seven times as active as toxin in ADPR-to-transferase 2 hook-up activity, (b) are nontoxic for animals and animal cells, and (c) are not inactivated by antitoxin. Thus, at this time one can only speculate as to the role of the subunit in the extreme toxicity of diphtherial toxin. There is as yet no evidence for a role of toxin in the synthesis of corynebacteriophages, and though tox$^+$ is a phage gene its relation to the biogenesis of toxin is unknown. The molecular configurations for which toxin shows specificity would seem just now to be the only clue to its possible mode of action in animal cells and perhaps to its origins from the lysogenic bacterial cell.

tox^+ AND THE TAXONOMY OF CORYNEBACTERIA

In the nineteen thirties the Leeds Group, under the inspiration of J. W. McLeod (36), showed that there was a correlation between the clinical severity of diphtheria and the colonial morphology of the organism isolated from the patient. On the basis of colonial morphology on chocolate tellurite agar, diphtheria bacilli could be assigned to one of three categories: (1) colonies having a rough appearance and associated with severe infections were designated *C. diphtheriae gravis;* (2) those of small size and a smooth appearance (dwarf, smooth colonies) and associated with less severe infections were designated *C. diphtheriae intermedius;* and (3) those having a smooth appearance and associated with mild infections were called *C. diphtheriae mitis.* The mode of selecting the Leeds types meant that they all carried the *tox+* gene. Of course, analogous colonial types lacking *tox+* exist (4), and they can be converted to toxinogeny through the agency of suitable bacteriophages.

Two of the three colonial "types," *gravis* and *mitis,* have shed some light on the possible origins and interrelationships of strains of *C. diphtheriae* and *C. ulcerans.* This insight comes from the following lines of evidence. Saragea and Maximescu (40, 55) have assembled a valuable set of corynebacteriophages which discriminate between smooth and rough strains of *C. diphtheriae.* With certain of these phages it is possible to separate bacteria which produce different receptors for bacteriophage: *m*-receptors and *r*-receptors. *Gravis* strains have *r*-receptors; *mitis* strains, *m*-receptors. *Corynebacterium ulcerans,* originally described from an outbreak of diphtheria epidemiologically associated with the consumption of milk from one herd of dairy cattle in New York State and later responsible for diphtheritic infections in Scandinavia (22) and more recently in Romania (40), possesses both *r*- and *m*-receptors. These and other genetic markers of *C. ulcerans* (2) suggest that it may be the ancestor of the strains more commonly involved in human infections.

The extent to which the capacity to express *tox+* should be taken as evidence of genetic relatedness among corynebacteria, of course, remains to be assessed. It is worth noting here, nonetheless, that *tox+*

can be expressed not only in suitably lysogenized strains of *C. diphtheriae* and *C. ulcerans* but in *C. ovis* as well.

SUBSUM

Tox$^+$, the gene governing the synthesis of diphtherial toxin, has been found in a wide variety of corynebacteriophages, some of them being morphologicaly, serologically, and genetically distinct. The expression of *tox$^+$* is maximal in diphtheria bacilli of the low-iron phenotype (LFC) and minimal in bacilli of the high-iron (HFC) phenotype. These phenotypes are morphologically distinct and exhibit certain differences in ultrastructure. Once toxin synthesis is turned on in LFC it can continue in the absence of bacterial DNA synthesis. Nontoxinogenic members of the three species of corynebacteria, *C. diphtheriae*, *C. ulcerans*, and *C. ovis*, all produce the same diphtherial toxin after lysogenization with suitable *tox$^+$* phages.

REFERENCES

1. Anderson, P. S., Jr., and P. B. Cowles (1958). Effect of antiphage serum on the virulence of *Corynebacterium diphtheriae, J. Bact.,* **76**, 272–280.
2. Arden, S. B., and L. Barksdale (1970). Nitrate reductases and the classification of corynebacteria infecting man and animals, *Bact. Proc.,* 239.
3. Barksdale, Lane (1955). Sur quelques bacteriophages de *Corynebacterium diphtheriae* et leurs hotes, *Compt. rend.,* **240**, 1831–1833.
4. Barksdale, L. (1959). Lysogenic conversions in bacteria, *Bact. Rev.,* **23**, 202–212.
5. Barksdale, L., L. Garmise, and K. Horibata (1960). Virulence, toxinogeny, and lysogeny in *Corynebacterium diphtheriae, Ann. N.Y. Acad. Sci.,* **88**, 1093–1108.
6. Barksdale, L., L. Garmise, and R. Rivera (1961). Toxinogeny in *Corynebacterium diphtheriae, J. Bact.,* **81**, 527–540.
7. Barksdale, W. L., and A. M. Pappenheimer, Jr. (1954). Phage-host relationships in toxigenic and nontoxigenic diphtheria bacilli, *J. Bact.,* **67**, 220–232.
8. Belsey, M. A., M. Sinclair, M. R. Roder, and D. R. LeBlanc (1969). *Corynebacterium diphtheriae* skin infections in Alabama and Louisiana: a factor in the epidemiology of diphtheria, *New Eng. J. Med.,* **280**, 135–141.
9. Bizzini, B., R. O. Prudhomme, A. Turpin, and M. Raynaud (1963). Essai de mise en évidence de liasons disulfure dans la toxine tétanique et la toxin diphtérique, *Bull. Soc. Chim. Biol.,* **45**, 925–932.

10. Collier, R. J. (1967). Effect of diphtheria toxin on protein synthesis: inactivation of one of the transfer factors, *J. Mol. Biol.*, **25**, 83–98.

11. Collier, R. J., and H. A. Cole (1969). Diphtheria toxin subunit active *in vitro*, *Science*, **164**, 1179–1182.

12. Collier, R. J., and A. M. Pappenheimer, Jr. (1964). Studies on the mode of action of diphtheria toxin. II. Effect of toxin on amino acid incorporation in cell-free systems, *J. Exp. Med.*, **120**, 1019–1039.

13. Collier, R. J., and J. A. Traugh (1969). Inactivation of aminoacyl transferase II by diphtheria toxin, *Cold Spring Harbor Symp. Quant. Biol.*, **34**, 589–594.

14. Edward, D. G., and V. D. Allison (1951). Diphtheria in immunized persons with observations on a diphtheria-like disease associated with non-toxigenic *C. diphtheriae*, *J. Hyg.*, **49**, 205–219.

15. Edwards, D. C., and P. A. Seamer (1960). The uptake of iron by *Corynebacterium diphtheriae* growing in submerged culture, *J. Gen. Microbiol.*, **22**, 705–712.

16. Freeman, V. J. (1951). Studies on the virulence of bacteriophage-infected strains of *C. diphtheriae*, *J. Bact.*, **61**, 675–678.

17. Goldzimer, E., S. B. Arden, and L. Barksdale (1968). Lysogeny, toxinogeny, and production of nitrate reductase by *Corynebacterium ulcerans*, *Bact. Proc.*, **V**, 94.

18. Goor, R. S. (1968). New form of diphtheria toxin, *Nature*, **217**, 1051–1053.

19. Goor, R. S., A. M. Pappenheimer, Jr., and E. Ames (1967). Studies on the mode of action of diphtheria toxin. V. Inhibition of peptide bond formation by toxin and NAD in cell-free systems and its reversal by nicotinamide, *J. Exp. Med.*, **126**, 923–939.

20. Groman, N. B. (1955). Evidence for the active role of bacteriophage in the conversion of nontoxigenic *Corynebacterium diphtheriae* to toxin production, *J. Bact.*, **69**, 9–15.

21. Groman, N. B. (1960). Conversion by bacteriophage, a factor in bacterial variation (Proceedings of the 21st Annual Biological Colloquium, Oregon State College), published in *Microbial Genetics*.

22. Gundersen, W. B., and S. D. Hendriksen (1959). Conversion in *Corynebacterium belfanti* by means of temperate phage originating from a toxigenic strain of *Corynebacterium diphtheriae*, type *mitis*, *Acta Path. Microbiol. Scan.*, **47**, 173–181.

23. Hirai, T., T. Uchida, Y. Shinmen, and M. Yoneda (1966). Toxin production by *Corynebacterium diphtheriae* under growth limiting conditions, *Biken's J.*, **9**, 19–31.

24. Holmes, R. K., and L. Barksdale (1969). Genetic analysis of *tox+* and *tox−* bacteriophages of *Corynebacterium diphtheriae*, *J. Virol.*, **3**, 586–598.

25. Holmes, R. K., and L. Barksdale (1970). Comparative studies with *tox+* and *tox−* bacteriophages, *J. Virol.*, **5**, 783–794.

26. Honjo, T., Y. Nishizuka, O. Hayaishi, and I. Kato (1968). Diphtheria toxin-dependent adenosine diphosphate ribosylation of aminoacyl transferase II and inhibition of protein synthesis, *J. Biol. Chem.*, **243**, 3553–3555.

27. Ipsen, J. (1946). Circulating antitoxin at the onset of diphtheria in 425 patients, *J. Immunol.*, **54**, 325–347.

28. Iskierko, J. (1965). Chemical composition of diphtheria toxin and toxoid. II. N-terminal amino acid of diphtheria toxin and toxoid, *Med. Dosw. Mikrobiol.*, 17, 217–223 (in Polish).
29. Kato, I., H. Nakamura, J. Uchida, and T. Katsura (1960). Purification of diphtheria toxin. II. The isolation of crystalline toxin protein and some of its properties, *Jap. J. Exp. Med.*, 30, 129–145.
30. Kato, I., and A. M. Pappenheimer, Jr. (1960). An early effect of diphtheria toxin on the metabolism of mammalian cells growing in culture, *J. Exp. Med.*, 112, 329–349.
31. Katsura, T., I. Kato, H. Nakamura, and J. Koyama (1957). Purification of diphtherial toxin, *Jap. J. Microbiol.*, 1, 213–234.
32. Klebs, E. (1883). Über Diphtherie, *Verh. Cong. inn. Med.*, 2, 139–154.
33. Lampidis, T., and L. Barksdale (1970). Restriction and modification of the phage of *Corynebacterium diphtheriae*, strain PW8, *Bact. Proc.*, 317.
34. Loeffler, F. (1884). Untersuchungen über die Bedeutung der Mikroorganismen für die Entstehung der Diphtherie beim Menschen, bei der Taube und beim Kalbe, *Mitt. klin. Gesundht.*, 2, 421–499.
35. Lwoff, A. (1953). Lysogeny, *Bact. Rev.*, 17, 269–337.
36. McLeod, J. W. (1943). The types *mitis, intermedius,* and *gravis* of *Corynebacterium diphtheriae, Bact. Rev.*, 7, 1–41.
37. Matsuda, M., and L. Barksdale (1966). Phage-directed synthesis of diphtherial toxin in nontoxinogenic *Corynebacterium diphtheriae, Nature*, 210, 911–913.
38. Matsuda, M., and L. Barksdale (1967). System for the investigation of the bacteriophage-directed synthesis of diphtherial toxin, *J. Bact.*, 93, 722–730.
39. Maximescu, P. (1968). New host-strains for the lysogenic *Corynebacterium diphtheriae* Park Williams No. 8 strain, *J. Gen. Microbiol.*, 53, 125–133.
40. Maximescu, P., A. Pop, A. Oprisan, and E. Potoral (1968). Relations biologique entre *Corynebacterium ulcerans, Corynebacterium ovis* et *Corynebacterium diphtheriae, Arch. Roum. Path. Exp. Microbiol.*, 27, 733–750.
41. Moriyama, T., and L. Barksdale (1967). Neuraminidase of *Corynebacterium diphtheriae. J. Bact.*, 94, 1565–1581.
42. Mueller, J. H. (1941). The influence of iron on the production of diphtheria toxin, *J. Immunol.*, 42, 343–351.
43. Mueller, J. H., and P. A. Miller (1941). Production of diphtheria toxin of high potency (100 L_f) on a reproducible medium, *J. Immunol.*, 40, 21–32.
44. Pappenheimer, A. M., Jr., H. P. Lundgren, and J. W. Williams (1940). Studies on the molecular weight of diphtheria toxin, antitoxin and their reaction products, *J. Exp. Med.*, 71, 247–262.
45. Park, W. H., and A. W. Williams (1896). The production of diphtheria toxin, *J. Exp. Med.*, 1, 164–185.
46. Pope, C. G., and M. F. Stevens (1953). Isolation of a crystalline protein from highly purified diphtheria toxin, *Lancet*, 265(2), 1190.
47. Pope, C. G., and M. Stevens (1958). The purification of diphtheria toxin and the isolation of crystalline toxin protein, *Brit. J. Exp. Path.*, 39, 139–149.
48. Ramon, G. (1923). Sur le pouvoir floculant et sur ses propriétés immunisantes d'une toxine diphtérique rendue anatoxique (anatoxine), *Compt. rend.*, 177, 1338–1340.

49. Raynaud, M., B. Bizzini, and E. Relyveld (1965). Composition en amino-acides de la toxine diphtérique purifiée. *Bull. Soc. Chim. Biol.*, 47, 261–266.
50. Relyveld, E. H., E. Henocq, and M. Raynaud (1962). Étude sur la sensibilisation aux antigens élaborés par diverses souches de corynebactéries diphtériques et non diphtériques. *Ann. Inst. Pasteur*, 103, 590–604.
51. Relyveld, E. H., and M. Raynaud (1964). Préparation de la toxine diphtérique pure cristallisée à partir de cultures en fermenteur. Caractères de la toxine pure, *Ann. Inst. Pasteur*, 107, 618–634.
52. Righelato, R. C. (1969). The distribution of iron in iron-deficient toxin-synthesizing and in excess-iron non-toxin-synthesizing *Corynebacterium diphtheriae*, *J. Gen. Microbiol.*, 58, 411–419.
53. Righelato, R. C., and P. A. van Hemert (1969). Growth and toxin synthesis in batch and chemostat cultures of *Corynebacterium diphtheriae*, *J. Gen. Microbiol.*, 58, 403–410.
54. Roux, E., and A. Yersin (1888). Contribution à l'étude de la diphtérie, *Ann. Inst. Pasteur*, 2, 629–661.
55. Saragea, A., and P. Maximescu (1966). Phage typing of *Corynebacterium diphtheriae*. Incidence of *C. diphtheriae* phage types in different countries, *Bull. World Health Org.*, 35, 681–689.
56. Strauss, N. (1960). The effect of diphtheria toxin on the metabolism of HeLa cells. II. The effect on nucleic acid metabolism, *J. Exp. Med.*, 112, 351–359.
57. Strauss, N., and E. D. Hendee (1959). The effect of diphtheria toxin on the metabolism of HeLa cells, *J. Exp. Med.*, 109, 144–163.
58. van Heyningen, W. E., and S. N. Arseculeratne (1964). Exotoxins, *Ann. Rev. Microbiol.*, 18, 195–216.
59. Warren, L., and C. W. Spearing (1963). Sialidase (neuraminidase) of *Corynebacterium diphtheriae*, *J. Bact.*, 86, 950–955.
60. Wilson, Sir Graham S. (1967). *The Hazards of Immunization*, University of London, The Athlone Press, pp. 39–42.
61. Yoneda, M., and H. Ishihara (1960). Studies on the iron-binding site of diphtheria bacilli. 1. Quantitative binding of iron by iron-deficient cells of a toxinogenic strain of *Corynebacterium diphtheriae*, *Biken's J.*, 3, 11–26.
62. Yoneda, M., and L. Barksdale. Unpublished data.

Cholera–an exotoxinosis rediscovered

W. E. VAN HEYNINGEN

This is an essay on cholera written—with great daring, in the modest hope of pleasing André Lwoff (Fig. 1)—by a newcomer to the field who himself has done hardly any experimental work on the subject.

For many years I taught my students that there were only two infectious diseases that we understood properly in the sense that we knew what weapon the infecting organism used in bringing about its harmful effects in the host. The two diseases I had in mind were diphtheria and tetanus. In these cases the mechanism of microbial pathogenic action is simple—the infected hosts are poisoned by extremely potent poisons (exotoxins) that are produced by the organisms. There is another exotoxinosis, botulism, but botulism is a food poisoning, not an infectious disease. Many other pathogenic bacteria produce exotoxins and endotoxins, some of them also extremely potent, but in these cases the part that the toxin plays is obscure and the diseases cannot—and should not—be considered as simple endo- or exotoxinoses. However, in recent years it has become clear that another ancient and terrible infectious disease is an exotoxinosis. This disease is cholera. It is interesting to consider why it has taken so long for the exotoxinotic nature of cholera to be revealed.

In 1884 Loeffler, reflecting on the fact that organs at a distance from the site of infection by the diphtheria bacillus were gravely affected, although sterile, suggested that the bacillus produced a poison which was carried throughout the body by the blood. Four years later Roux and Yersin isolated such a poison from a laboratory culture of the

Figure 1. André Lwoff in Oxford, about to be given the honorary degree of Doctor of Science at Encaenia on 6 June 1959.

diphtheria bacillus. In 1885 Nicolaier made a similar suggestion about the tetanus bacillus, an organism which produces frightening effects in the host despite its incapacity to establish more than an extremely circumscribed bridgehead at the site of infection. In 1890 Faber confirmed the existence of tetanus toxin, and in 1891 von Behring and Kitasato showed that both diphtheria and tetanus toxins elicited antitoxins in the host that were capable of neutralising the toxins. Herein lay the means of protecting people against these formidable infectious diseases: by immunizing them in advance against the toxins, which do all the damage, it is possible to protect them against subsequent attacks of the diseases. In most developed countries diphtheria and tetanus have now been practically wiped out.

Cholera had raged in India, particularly about the delta of the Ganges in Bengal, during the eighteenth century and before, but in the nineteenth century it broke out of India and spread throughout the world in five great pandemics. The first (1817) started with a violent epidemic in India, spread eastwards as far as China and Japan, westwards to the Caspian Sea. The second (1829) spread eastwards again to China, westwards through the continent of Europe to England, and across the ocean to the Pacific Coasts of North and South America. The third (1852) and fourth (1863) pandemics also spread as far as the New World. The fifth (1887) was investigated by Robert Koch in Egypt and Calcutta; he proved that the disease was the result of gastro-intestinal infection by the comma bacillus which we now know as *Vibrio cholerae.*

A sixth pandemic, lasting the first quarter of this century, killed over 300,000 people in European Russia alone. That was essentially the last outbreak of cholera in the West, but in 1958 there was a seventh pandemic, spreading from Southeast Asia and the Western Pacific through the Middle East to the borders of southeastern Europe. The disease continues almost unabated in East Pakistan and West Bengal; since 1900 the annual deaths from cholera in India and Pakistan have fluctuated from 0.06 million to 0.8 million, exceeding 0.25 million on 26 occasions, the last being 1945.

Cholera is a truly dreadful disease. The victim suffers from sudden, explosive, unremitting diarrhoea, often accompanied by vomiting, and may die within hours of the onset of the disease. His stools are watery, only slightly cloudy, with a faintly sweet odour. In 1832 the Assistant Surgeon of the United States Army described a cholera patient in these terms:

The face was sunken, as if wasted by lingering consumption; perfectly angular, and rendered peculiarly ghastly by the complete removal of all the soft solids, and their places supplied by dark lead-coloured lines; the hands and feet were bluish white, wrinkled as when long macerated in cold water; the eyes had fallen to the bottom of their orbs, and evinced a glaring vitality, but without mobility; and the surface of the body was cold.

This condition results from the fact that the victim of cholera loses in his watery stools and in his vomit an enormous volume of fluid—it may be as much as 30 litres a day. This fluid is isotonic, its composition approximating that of an ultrafiltrate of blood. Since the daily intake of water is nearly zero in cholera, the bulk of the fluid lost every day must come from the water in the body. In a grown man this totals about 50 litres, of which about 3.5 litres circulate in the blood. The lost water must be derived via the blood from the interstitial and intracellular water of the body. Consequently the victim of cholera becomes drastically dehydrated. John Snow[1] (1855) described the result as follows:

The loss of water from the blood causes it to assume a thick tarry appearance, so well known to all who have opened a vein in cholera. The diminished volume of the blood causes many of the symptoms of a true haemorrhage, as debility, faintness, and coldness; while these effects are much increased by its thick and tenacious condition, which impedes its passage through the pulmonary capillaries, thereby reducing the contents of the arteries throughout the system to the smallest possible amount, as indicated by the small thready pulse. The interruption to the pulmonary circulation occasioned by the want of fluidity of the blood, is the cause of the distressing feeling of want of breath.

How does the cholera vibrio cause the remarkably rapid flow of fluid into the gut at the expense of the blood and the tissues of the body? The great pathologist Virchow, on the basis of epithelial slough observed on autopsies of victims who had been in prolonged shock, ascribed the fluid losses in cholera to the destruction and denudation

[1] John Snow was the tough physician who put an end to "the most terrible outbreak of Cholera which ever occurred in this kingdom" in the Soho district of London in 1854 by persuading the Board of Guardians of St James's Parish to remove the handle of the water pump in Broad Street.

of epithelium and the loss of plasma. His views were widely accepted and perpetuated (and in more recent times modernised by the suggestion that the mucinase of the cholera vibrio is responsible for the stripping of the mucosa; see Burnet and Stone, 1947). However, Goodpasture (1923) showed that the mucosa retained its integrity in cholera, and this was confirmed by Gangaraosa et al. (1960) by means of small intestinal biopsies of cholera patients; Weaver, Johnson, and Phillips (1948) found that cholera stools were much lower in protein content than plasma; Gordon (1962) reported that macromolecules injected into the bloodstream during cholera did not pass into the gut; and Finkelstein et al. (1966) and others disproved the mucinase hypothesis (see also Miles, 1955, on the fallacy of the appropriate substrate).

On 26 July 1884 Robert Koch addressed a conference in the Imperial German Public Health Department in Berlin, assembled to discuss the cholera problem and presided over by Virchow. Koch was reporting on the finding he had just made during his investigation of the cholera outbreak in Egypt that the disease was caused by *Vibrio cholerae*. He noted that the bacilli were confined to the intestine and did not spread into the blood or even the mesenteric glands. On postmortem examination there was no visible damage to the walls of the intestine (*pace* Virchow) or any other tissue. "How does it happen, then," he asked, "that the vegetation of bacteria in the intestine can cause death?" His answer was that the bacillus produced a "specific" poison "that acted upon the epithelium."

This was, it should be reiterated, in 1884, the year that Loeffler postulated diphtheria toxin, and the year before Nicolaier postulated tetanus toxin. The existence of diphtheria and tetanus toxins was confirmed a ycar or two later, but recognition of the existence of cholera exotoxin was delayed for 75 years. Why? The answer is that the demonstration of diphtheria and tetanus exotoxins in culture filtrates of the parent bacilli had been a simple matter—parenteral injection of small volumes of the filtrates into experimental animals killed them. But injection of comparatively large volumes of cholera culture filtrates by the normal parenteral routes (intravenous, intramuscular, subcutaneous, intraperitoneal) had produced no observable effect. It is therefore no wonder that Koch's suggestion of cholera poisoning was abandoned.

In the apparent absence of an exotoxin, all kinds of suggestions were made to account for the action of cholera vibrio in the gut, and the one that gained the most support—and created the greatest confusion—was the hypothesis, or rather the assumption, that the *endo*toxin of Boivin et al. (1934) was responsible. The endotoxins are antigenic complexes of polysaccharide, phospholipid and protein extractable from the cell bodies, not from fresh culture filtrates, of a number of Gram-negative organisms, including *Vibrio cholerae*. They are moderately toxic, and they all produce the same symptoms when introduced parenterally into experimental animals, regardless of the parent organisms. Many authorities believed, as did Wilson and Miles (1946), that "the cholera vibrio does not secrete a true soluble exotoxin but . . . contains endotoxins which are liberated on the autolysis of the bacilli in culture or on the active disintegration of the bacilli by the cells of the animal body."

Much of the great amount of work that has been done on the assumption that endotoxin plays an important part in the diarrhoea of cholera is uninterpretable, largely because ill-defined preparations of endotoxin have been used. However, it is now doubted whether parenterally active endotoxic principles derived from bacterial cell bodies play such an important part in cholera as was first thought. This change in attitude has occurred because an enterally active exotoxin has been demonstrated in filtrates of young cultures of *Vibrio cholerae*.

In 1959, 75 years after Koch's original suggestion of a cholera exotoxin, the existence of such a toxin was demonstrated at the Medical College of Calcutta by De and his colleagues (De, 1959; De, Ghose, and Sen, 1960; De, Ghose, and Chandra, 1962). They found that, when sterile culture filtrates of the cholera vibrio (which produced, as usual, no effect when introduced parenterally) were injected into ligated loops of the small intestine of the rabbit, there was an accumulation of fluid in the gut loop. Such an accumulation of fluid took place only after injection of the extracellular fluid of vibrio cultures, and not after injection of an (endotoxin-rich) extract of ultrasonically disintegrated washed bacilli. Moreover, it appeared with the filtrates of young cultures. This was the first serious demonstration of an exotoxin of *Vibrio cholerae* that produced in the gut the same symptoms that appear in the natural disease. The fact that the toxin is almost without effect when introduced parenterally, and is active only when introduced directly into the lumen of the gut, accounts for the long time required for its existence to be recognized.[2]

The presence of cholera exotoxin has been demonstrated in the stools of cholera patients by Craig (1965). Craig (1965, 1966) also showed that, when dilutions of cholera culture filtrates and cholera stools are injected into the skin of rabbits and guinea pigs, the blood vessels become permeable to protein and an induration is formed in the skin, which can easily be studied if a dye such as pontamine blue is injected intravenously shortly before the skin lesions are examined.

There has been much discussion about whether or not the gut-reactive and skin-reactive factors are identical (see Burrows, 1968), but recently this question appears to have been resolved. Mosley, Aziz, and Ahmed (1970) at the Pakistan-SEATO Cholera Research Laboratory in Dacca investigated the question by applying an immunological principle that had been well tried in the field of bacterial exotoxins. They titrated a number of antitoxic sera from different species against a test cholera exotoxin, using in each case both the gut-loop test and the skin test as indicators of toxin excess over antitoxin, and found for each serum the same titre with both tests. It is known that when any two antigens have been introduced simultaneously into each of two animals the chance that the homologous antibodies have been produced in the same ratio in both animals is extremely small, and for this to happen with a number of animals is practically impossible. Therefore, the only reasonable interpretation to place on Mosley's results is that the gut and skin activities in his preparation of exotoxin were due to the same toxin. (This does not exclude the possibility that other cholera exotoxin preparations may contain a skin-reacting factor with no gut activity, and vice versa, but such factors have not yet been demonstrated.) The vascular permeability reaction of cholera toxin is important because it offers a much simpler way of assaying the toxin, and it may provide clues to the action of the toxin in the intestine.

Cholera exotoxin has been purified and prepared in a state approaching immunological and kinetic homogeneity (Finkelstein and Lo-Spalluto, 1969; Richardson, 1970; Spyrides, 1970). The purified exo-

[2] For the first explicit (and lucid) assessment of the part played by the exotoxin in cholera, see MacLeod (1965).

toxin preparation of Finkelstein and LoSpalluto ("choleragen") had an approximate molecular weight of 61,000, as estimated by Sephadex gel and membrane filtration; 0.4 ng caused an induration when injected into the rabbit skin, 0.25 μg produced diarrhoea when fed to infant rabbits, and 0.4 μg caused fluid accumulation when introduced into ligated gut loops in adult rabbits (Finkelstein, 1969). Finkelstein and LoSpalluto (1969) also isolated a protein ("choleragenoid") which was antigenically identical with "choleragen" but had an estimated molecular weight of approximately 42,000 and only about one-thousandth the biological activity of "choleragen" (which may have been due to contamination with the latter).

The pathophysiology of the action of cholera exotoxin, that is, the way in which it causes an accumulation of isotonic fluid in the small intestine, is of course a matter of great interest. In view of the enhancing effect of cholera toxin on vascular permeability in the skin it would not be unreasonable to consider the possibility of a filtration of fluid from the network of capillaries that is present in the intestinal villae. Since in the skin test blood vessels become permeable to protein under the influence of the exotoxin, it would be necessary to postulate a blockage of the passage of protein by the intact epithelium, with some resultant oedema and a return of the protein to the blood via the lymphatic system (see Finkelstein, 1969). However, Carpenter et al. (1968) were unable to demonstrate an increase in the protein content of thoracic duct lymph during diarrhoea, and more recently they showed that the rate of production of gut fluid during experimental cholera in dogs was independent of mesenteric blood flow and pressure over a wide range (Carpenter and Greenough, 1968; Carpenter, Greenough, and Sack, 1969). This makes it unlikely that filtration of fluid from blood to gut lumen is an important mechanism of fluid loss in cholera. The action of cholera exotoxin that is common to its effect on blood vessels in the skin and on the lining of the intestine has not yet been discovered.

It can be calculated from the available data on unidirectional flux of sodium ions and water (e.g., Berger, 1960; Soergel, Whalen, and Harris, 1968) that in normal man the volume of water that flows through the cells of the gut epithelium from the blood to the lumen and back again to the blood is at least about 100 litres per day (apart from about 10 litres of ingested water, saliva, and other secretions).

With such a large volume of water surging about, only a small change in the movement in one direction or the other would need to be induced by cholera toxin to account for the accumulation of up to 30 litres of fluid per day in the gut. In recent years the evidence gained by workers in various laboratories (see Greenough and Carpenter, 1969) has seemed to suggest that the toxin causes an increased flow from blood to lumen without affecting the flow in the opposite direction. However, work at present in hand (as yet unpublished) suggests that in cholera the flow of water is reduced in both directions, but that the flow from lumen to blood is decreased to a greater degree than that in the opposite direction. The outcome of this is a reduction in the total flux of water and a net accumulation of water in the gut.

There appear to be several mutually independent mechanisms for the transport of sodium ions and water through the gut epithelium, one of which is stimulated by glucose and another by glycine. The latter two mechanisms do not appear to be affected in cholera patients (Hirschhorn et al., 1968; Pierce et al., 1968) or in toxin-treated Thiry-Vella loops in the small intestine of dogs (Iber et al., 1968).

Recently studies have been made of the effect of cholera exotoxin on isolated biological systems: crude and highly purified preparations of cholera exotoxin have been found to have observable effects on ion transport through sheets of gut epithelium clamped between two chambers (see Greenough and Carpenter, 1969), and to stimulate the production of glycerol in suspensions of rat epididymal lipocytes (Greenough, Vaughan, and Pierce, 1970). It is possible that work on these lines may provide clues to possible ways of studying the action of this toxin at the molecular level.

We have seen that it is possible to protect people effectively against diphtheria and tetanus by actively immunizing them against the exotoxins produced by the organisms causing the diseases. This active immunity, which results from the injection of two doses of diphtheria or tetanus toxoids, followed sometimes by a third injection 6 months later, protects nearly all the individuals so treated, and this immunity lasts for many years, if not for life. Should it not also be possible to immunize people effectively and lastingly against cholera by immunizing them against the exotoxin? Although the best of the cholera vaccines currently used apparently confer immunity to cholera on a high proportion (85%) of those vaccinated, this protection does not last longer than 6 months (Oseasohn, Benenson, and Fahimuddin, 1965). These cholera vaccines are prepared from whole cells and contain very few, if any, of the extracellular products of the cholera vibrio. They are designed to produce antibacterial immunity, rather than anti-exotoxic immunity. They were developed before the important role of exotoxin was recognized, during the period when the importance of endotoxin was overestimated.

Kasai and Burrows (1966) and Finkelstein and Atthasampunna (1967) showed that immunization by parenteral injection of cholera exotoxin in rabbits protected ligated gut loops in these animals against fluid accumulation induced by subsequent intraluminal exposure to cholera exotoxin, and Craig (1967) demonstrated that immunization with formalinized exotoxoid protected dogs against subsequent challenge with crude cholera culture filtrates. Curlin et al. (1968) showed that intramuscular injection of crude cholera exotoxin into dogs produced antitoxin in the serum of their subjects, and that animals so treated were immune to intraluminal challenge both with exotoxin and with whole living cultures. The protection conferred upon the dogs in this way was more effective than that conferred by commercial vaccine.

The comparison of cholera with diphtheria and tetanus, in the context of the possibility of immunizing against cholera by immunizing against the exotoxin, may, however, be false. In the case of diphtheria and tetanus, the state of immunity is induced by parenteral administration of toxoid and the level of circulating (humoral) antitoxin is generally regarded as a good indication of the degree of immunity, since the challenge by the toxin in the natural infection takes place in the tissues, that is, in a parenteral situation. In cholera, on the other hand, the organism grows in the gut; the toxin is produced in the gut and, as far as we know, is fixed on the gut epithelium and acts upon it.

It would be reasonable to ask whether humoral antitoxin, as induced by parenteral immunization, would protect the wall of the gut, and whether the antitoxin should not actually be present in the gut wall to be effective. Curlin and Carpenter (1970) have recently shown that circulating humoral antitoxin does indeed protect dogs against intraluminal challenge with cholera toxin. Dogs that had, and others that had not, been actively immunized against the toxin had their blood supplies to isolated loops of gut cross-circulated in such a way that an isolated loop of gut from the immunized dog received its blood circulation from the nonimmunized animal, and vice versa. In each dog another isolated loop of gut continued to receive its blood supply directly. The various gut loops were challenged intraluminally with toxin: regardless of the immunization of the dogs, gut loops receiving nonantitoxic blood produced more fluid in response to intraluminal toxin challenge; those receiving antitoxic blood failed to produce more fluid in response to intraluminal toxin challenge.

This experiment suggests that humoral, parenterally induced antitoxin is capable of neutralising enterally presented toxin, and encourages the idea of the prevention of cholera by active immunization against the exotoxin. Indeed, parenteral active immunization is likely to be more effective than the immunization resulting after survival from the natural disease. The situation with cholera may be similar to that with tetanus, and unlike that with diphtheria. In the course of the natural disease of diphtheria the surviving patient is exposed parenterally to enough toxin, for a long enough period, to be immunized, probably for life, against a recurrence of the disease. With tetanus the situation is different: tetanus toxin is so potent that the lethal dose is not enough to confer even measurable immunity, let alone effective immunity, and therefore the survivor of an attack of tetanus, who by definition has been exposed to less than a lethal dose of toxin, has not been immunized against the disease. Indeed there have been several cases of recurring tetanus in man in the United States, rare though the disease is in that country.

People surviving cholera have measurable antitoxin titres against the exotoxin. We have no evidence, however, that such persons have been exposed to enough toxin, for a long enough period, to produce effective immunity, and in any case the toxin has been presented by the enteral route, which is very inefficient for immunization. The levels of antitoxin that are likely to be obtained after the parenteral administration of large doses of cholera toxoid (which is actually more immunogenic than the toxin: Craig, 1967; Feeley and Roberts, 1969), properly spaced and enhanced by an adjuvant, are likely to be orders-of-magnitude higher than those observed in people convalescent from cholera. It may well be that such levels of immunization are necessary to protect against the disease. Moreover, there is a suggestion that the presence of exotoxin in the gut enhances the capacity of *Vibrio cholerae* to produce cholera (Sack and Carpenter, 1969). It is possible that the isotonic fluid poured into the gut in response to the toxin produces a favourable medium for further growth of the vibrio. Thus active immunization against the toxin, by stopping the response of the gut to the toxin, may have an antibacterial as well as an antitoxic effect.

A possible difficulty with active immunization against cholera may arise from the short incubation period and rapid course of the disease. Individuals who are actively immunized against diphtheria or tetanus are protected perhaps not so much by the actual level of antitoxin circulating at the time of infection as by the secondary production of antitoxin in response to the stimulus of infection; in cholera there may not be time enough for such an anamnestic response to come into play.

Clearly there is a need for a large-scale field trial of the practical efficacy of active immunization against cholera by means of cholera toxoid.

REFERENCES

Berger, E. Y. (1960). In *Mineral Metabolism,* edited by C. L. Comar, London, Academic Press, p. 249.

Boivin, A., I. Mesrobeanu, L. Mesrobeanu, and B. Nestorescu (1934). *Compt. rend. Soc. Biol. (Paris)*, 115, 306.

Burnet, F. M., and J. D. Stone (1947). *Austral. J. Exp. Biol. Med. Sci.*, 25, 219.

Burrows, W. (1968). *Ann. Rev. Microbiol.*, 22, 245.

Carpenter, C. C. J., and W. B. Greenough (1968). *J. Clin. Invest.*, 47, 15a.

Carpenter, C. C. J., W. B. Greenough, and R. B. Sack (1969). *J. Infect. Dis.*, 119, 182.

Carpenter, C. C. J., R. B. Sack, J. C. Feeley, and R. W. Steenberg (1968). *J. Clin. Invest.*, 47, 1210.

Craig, J. P. (1965). *Nature*, 207, 614.

Craig, J. P. (1966). *J. Bact.*, 92, 793.

Craig, J. P. (1967). *Symposium on Cholera, U.S.-Japan Cooperative Medical Research Program, Palo Alto,* p. 47.

Curlin, G. T., and C. C. J. Carpenter (1970). *J. Infect. Dis.*, in press.

Curlin, G. T., A. Subong, J. P. Craig, and C. C. J. Carpenter (1968). *Trans. Assoc. Amer. Physicians*, 81, 314.

De, S. N. (1959). *Nature*, 183, 1533.

De, S. N., M. L. Ghose, and J. Chandra (1962). *Trans. Roy. Soc. Trop. Med. Hyg.*, 56, 241.

De, S. N., M. L. Ghose, and A. Sen (1960). *J. Path. Bact.*, 79, 373.

Feeley, J. C., and C. O. Roberts (1969). *Texas Rep. Biol. Med.,* **27**, 213.

Finkelstein, R. A. (1969). *Texas Rep. Biol. Med.,* **27**, 181.

Finkelstein, R. A., and P. Atthasampunna (1967). *Proc. Soc. Exp. Biol. Med.,* **125**, 465.

Finkelstein, R. A., P. Atthasampunna, M. Chulasamaya, and P. Charunmethee (1966). *J. Immunol.,* **96**, 440.

Finkelstein, R. A., and J. J. LoSpalluto (1969). *J. Exp. Med.,* **130**, 185.

Finkelstein, R. A., P. Z. Sobocinski, P. Atthasampunna, and P. Charunmethee (1966). *J. Immunol.,* **97**, 25.

Gangarosa, F. T., W. R. Beisel, C. Benyajati, H. Sprinz, and P. Piyaratn (1960). *Amer. J. Trop. Med. Hyg.,* **9**, 125.

Goodpasture, E. W. (1923). *Philippine J. Sci.,* Sect. B, **22**, 413.

Gordon, R. S. (1962). *Proceedings of the SEATO Conference on Cholera,* p. 54.

Greenough, W. B., and C. C. J. Carpenter (1969). *Texas Rep. Biol. Med.,* **27**, 203.

Greenough, W. B., M. Vaughan, and N. F. Pierce (1970). *J. Infect. Dis.,* in press.

Hirschhorn, N. J., J. L. Kinzie, D. B. Sachar, R. S. Northrup, and R. A. Phillips (1968). *New Eng. J. Med.,* **279**, 176.

Iber, F. L., T. J. McGonagle, H. A. Serebro, E. H. Luebbers, T. M. Bayless, and T. R. Hendrix (1968). *J. Clin. Invest.,* **47**, 50a.

Kasai, G. J., and W. Burrows (1966). *J. Infect. Dis.,* **116**, 606.

Koch, R. (1884). *Deutsche med. Wochnschr.,* **32A**, 519.

MacLeod, C. M. (1965). *Proceedings of the Cholera Research Symposium, Honolulu,* Washington, D.C., U.S. Government Printing Office, p. 394.

Miles, A. A. (1955). In *Mechanisms of Microbial Pathogenicity,* Cambridge, Cambridge University Press, p. 13.

Mosley, W. H., K. M. S. Aziz, and A. Ahmed (1970). *J. Infect. Dis.,* in press.

Oseasohn, R. D., A. S. Benenson, and M. D. Fahimuddin (1965). *Lancet,* **288**, 450.

Pierce, N. F., J. G. Banwell, R. C. Mitra, G. J. Caranasos, R. I. Keimowitz, A. Mondal, and P. M. Manji (1968). *Gastroenterology,* **55**, 333.

Richardson, S. H. (1970). *J. Infect. Dis.,* in press.

Sack, R. B., and C. C. J. Carpenter (1969). *J. Infect. Dis.,* **119**, 150.

Snow, J. (1855). *On the Mode of Communication of Cholera,* London, John Churchill.

Soergel, K. H., G. E. Whalen, and J. A. Harris (1968). *J. Appl. Physiol.,* **24**, 40.

Spyrides, G. (1970). *J. Infect. Dis.,* in press.

Weaver, R. T., M. T. Johnson, and R. A. Phillips (1948). *J. Egypt. Public Health Assoc.,* **23**, 5.

Wilson, G. S., and A. A. Miles (1946). *Topley and Wilson's Principles of Bacteriology and Immunity,* 3rd Ed., London, Edward Arnold.

Going around in circles

S. FAZEKAS DE ST. GROTH

Homage to a great man is best paid in his own coinage: witness the intent of this volume. But André Lwoff is no believer in currency regulations and has never been one to let his field of work narrow his field of interest or influence, as all of us know: witness the scope of this volume. It is with an easy conscience, therefore, that I have chosen antigenic variation as my theme.

In looking for examples of antigenic variation, one might settle for some bacterium or other. But in these days of antibiotics microorganisms are dealt with by finding a chink in their metabolic armour; what antigens they may sport is all but irrelevant, and abiding interest in these problems places one at the ivory tower fringe of immunology.

On the whole, we are not much better off with viruses. Most of them seem antigenically stable, and the general principles of developing effective vaccines have been clear since the days of Pasteur. There are many jobs to be done in this area and even more people qualified to do them. Inexorably, further strains of virus will be attenuated, and yet others grown to titre and made safe as killed vaccines. But these projects are barren of adventure, even though they are fiercely competitive as is everything that leads to a marketable product. Abiding interest in this field places one at the treadmill fringe of immunology. So we are left with a group of viruses that has consistently got the better of us in the past, pays no heed to our antibiotics, and is as likely as not to turn up in new antigenic forms as soon as we develop a vaccine against last year's epidemic strain. The viruses of influenza are the textbook example of this group, and when dealing with antigenic structure I shall take all my examples from among them.

Let us see then what sort of evidence there is for antigenic variation amongst these viruses, and how good this evidence is. We take a number of strains from successive epidemics, and the corresponding antisera. When testing such sera against a range of antigens, we usually find that each is most effective at neutralizing its own virus, that is, gives the highest titre in the homologous combination. But when it comes to cross reactions, we are on shifting ground: some crosses almost equal the homologous titre in one direction, whereas the reciprocal test shows rather poor neutralization. It is noteworthy that this kind of thing largely follows the historical sequence: cross reactions are effective retrospectively but not prospectively. At this point one might feel tempted to make the facile observation that this is *why* a particular virus is able to establish itself in a population with reasonable immunity against its predecessors. But we should rather ask *how* such a thing could come about, and the first intelligent answer suggested that this might be so because viruses were antigenic mosaics.

Thus, in general, virus A would have antigen a on its surface (it is not necessary that it should have only a), virus B both a and b, and virus C three antigens, a, b, and c. This immediately suggests an experiment: if anti-C serum is absorbed with A, anti-B and anti-C antibodies should be left behind; with B virus as absorbent only anti-C should be left behind; and if one absorbs with C, nothing should be left behind. Such experiments have been performed with fixed doses of virus and, by and large, have given the expected answers. The few exceptions did not look serious enough to bother anybody, and the theory of antigenic mosaics became the vogue.

This theory is a comfort to epidemiologists: it acknowledges the fact that we don't know too much about what's going on (and the complexity of the virus is a good excuse for this ignorance), but if we persevere long enough the predominant antigens should run full circle and our labours shall bear fruit when the old antigens return and find all students of epidemiology well prepared for them. To the routine serologist this theory is a positive boon: it ensures that an unlimited number of laboratories will occupy themselves with an inexhaustible set of cross reactions, at the cost of minimal intellectual effort. Even the theoretical immunologist is better off, since he has an explicit hypothesis asking to be tested.

Now, what sort of tests would our immunologist do? Being a theoretical immunologist, he wouldn't do any, to begin with. He would just sit and muse over the fact that these viruses have RNA as their genetic material, and not too much of it either. At most, it could provide information for a protein of about 200,000 molecular weight or, say, five proteins of 40,000 molecular weight each. And antigenic mosaics have been postulated to be made up of at least 17 different components.

Since this sort of paradox cannot be resolved by meditation, our immunologist would be driven to design a method of counting how many antigenic sites there are on a virus particle, and of determining whether or not these are all of the same kind. Such tests have been devised and have been performed by now on a large number of strains. The answer is this: first, for a virus and homologous antibody, the number of antigenic sites comes to about 2000; then, for the same virus and heterologous antibody, the number of antigenic sites once again comes to 2000; and if we go on and take half-a-dozen cross-reacting combinations, the number of sites comes to 2000 in every case. This means either that there are at least six times 2000 sites on our virus, or that there are only 2000 sites and each of the different antibodies reacts with the same site, the combinations being of different firmness. Since it so happens that the experiment which determines the number of combining sites also measures the firmness of combination, our immunologist extracts this information, too, from his data. He finds that cross reactions are, indeed, of different firmness, an antiserum combining avidly with 2000 sites of the homologous antigen and less avidly with 2000 sites of related antigens.

The number 2000 is reasonable: if we work out how many antibody molecules can be accommodated on the surface of a virus at close packing, the number comes to about 2800. Thus under no circumstances could we have six times 2000, whereas 2000 would fit snugly. And we can go even one step further and ask our biochemist friends to take the virus to pieces, find out which is the antigenic component, and tell us how many of these there are on one particle.

The odds are that the biochemists won't be our friends by the time they're half through with this business, because it would be a hard task even if they had more than a few milligrammes of virus to play around with. Still, the work has been done, and we know now that there is only one viral component that binds neutralizing antibody, that this component is a homogeneous repeating unit, and that about 2000 such units are packed into the small rods jutting out of the viral envelope.

In the face of evidence like this, the theory of antigenic mosaics becomes of historical interest only. But we are still left with the fact of lopsided serologic crossing, and in need of some model to account for it. Such a model can be built on a single premise, and one that has been amply proven in practice, namely, that the response to any antigen is heterogeneous. Thus our antigen (to be imagined as a small area on a protein molecule, made up of not more than perhaps a dozen amino acid side chains) will stimulate one class of antibodies that fit it perfectly, and several classes that fit it only partially. When such a mixture of antibodies is tested against a related antigen differing by a single amino acid only (valine instead of alanine, for instance), all subclasses complementary at the alanine locus will be held at arm's length by valine and thus excluded from cross reactions. Antibodies to the second antigen, on the other hand, whether or not they contain regions complementary to valine, will encounter no steric hindrance: anything that can take in the bulky side chain of valine will not be bothered by sundry undersized amino acids at that spot.

Thus antigenic relationship cannot have a simple linear measure, but must be regarded as a vector. Its magnitude is defined by the number of elements shared between two determinants, and its direction by the nature of the different elements. In these terms our two antigens are *closely related* (they differ by a single element only), and the one containing valine is *senior* to the one containing alanine. As a consequence, where we postulated a mosaic of discrete antigens to account for lopsided cross reactions, we now consider predictably varying van der Waals contours and the steric exclusion of large side chains by cavities fitted to smaller ones.

Once again we have a testable hypothesis. By performing somewhat more elaborate cross-absorption tests with viruses ranked in order of seniority (in fact, with our old friends, viruses A, B, and C) it is easy to show that asymmetry of crossing rests on the exclusion of certain subpopulations of antibodies in junior sera by more senior antigens.

In a more critical experiment, a junior virus is grown in the presence of its antiserum or, better, the most avidly combining fraction of that serum. If conditions are so arranged that about 5 particles out of 10^9 grow through, we find that all of these are antigenic mutants, and all are senior to the parental strain. This proves that selection by antibody is for seniority. Repeating the same experiment with the most senior virus of our set, we find that it is much more difficult to isolate mutants, that the majority of the isolates carry the parental antigen and differ only in their affinity to the host cell, and that the few true antigenic mutants are distantly related to the parental antigen. On our model these mutants would be interpreted as differing by more than one locus within the determinant, that is, as being double mutants.

The contrast between the ease of selecting mutants from junior antigens and the difficulty of doing the same from senior ones is striking. Each sequence of mutational steps leads to a dead end; this has been the experience of all who have done similar experiments. Such a mechanism, if it operated in nature at all, could not account for the continued variability of these viruses. So we are forced to step outside our circle of closely related antigens and to look into the family relations within the whole known range of influenza A. The results, a large and thoroughly indigestible matrix, can be summarized by plotting the seniority of epidemic strains against the time of their emergence (Fig. 1).

This family of viruses falls into four groups. The four subtypes cross-react weakly with one another, but each is made up of closely related strains of virus. And the trend within the groups is far from random: the strains climb the ladder of seniority in much the same way as they did when we selected senior mutants under the pressure of antibody. Transition between subtypes seems abrupt and would remain a puzzle without the split trend in the A_2 era. Towards the middle of this period strains appeared, first in New Zealand and then in Australia, which differed from isolates over the rest of the world. They gave more distant cross reactions than is usual within a subtype and, on closer analysis, were found to differ from the reigning strains at two loci. The second of these two turned out to be the mutational locus characteristic of Hong Kong influenza, the pandemic strain of 1968–1969. Such *bridging antigens,* arising from junior members of one subtype and anticipating the next, are the link ensuring continuity, indeed, cyclic arrangement

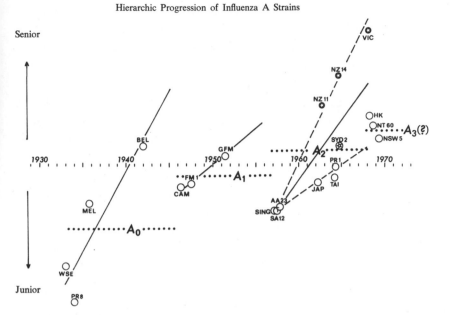

Hierarchic Progression of Influenza A Strains

Figure 1. *The ranking order of influenza A viruses*. Dotted lines represent the mean hierarchic level and span of the subtypes; continuous lines, the trend of seniority within a subtype. The starred symbols mark the "bridging strains" of subtype A_2; the divergent trend within this subtype is shown by broken lines.

of subtypes. Their antigenic pattern would be, by our definitions, senior to both parental and filial subtypes, and there is perhaps no better evidence for this than the fact that the region where these bridging strains arose was the only one that escaped the pandemic of 1969.

The cyclic nature of antigenic variation is, of course, not proven, but there are straws in the wind. By dint of good fortune, influenza is amenable to palaeovirological research: the viruses of this disease are closely enough related to render each successive exposure a cross stimulation of the primary immunity. Thus, irrespective of the vaccinating antigen, each cohort responds by producing antibodies against the strains prevalent in its earliest childhood. The nature of these antibodies, in its turn, defines the subtype prevalent in that era. Such investigations have by now covered large populations on three continents, and the results show unequivocally that the subtype we call A_2 was also present between

1889 and 1899, while viruses similar to the Hong Kong strain (sub-type A_3) arose around the turn of the century. There is a gap of information between 1910 and 1933, allowing for two, perhaps three, more subtypes, making six or seven in all.

Variation in viruses rests on variation in their genetic material. But changes in nucleotide sequence must greatly exceed the frequency of acceptable variants. Apart from the trivial reason, the redundancy of the code, this is so because each mutation challenges the structural and functional integrity of the virion. Thus the architecture of the simpler viruses, in which the protein shell is both the antigenic and the sole structural unit, will not tolerate antigenic changes unless they leave the clathrate structure untouched. Such mutations will be rare and, in all likelihood, conservative substitutions. Soon in evolutionary time the most advantageous (most senior?) forms will be fixed and remain predominant, since natural selection, largely in the form of herd immunity in the case of animal viruses, keeps screening any changes.

At the other end of the scale are viruses whose antigenic determinants form no part of a structural unit and have hence greater initial freedom to vary. If such a determinant is open to simultaneous substitutions at several loci, it will have no subtypes and will, under the pressure of antibody, end up in the most senior pattern within its conformational limits. The result is once again a fixed ground state, as any further change would be selectively neutral or positively disadvantageous.

Between these extremes lie structures with peculiar constraints on conformation. Influenza A viruses will accept, it appears, sequential substitutions at any one of a number of amino acid loci within the antigenic area, while the simultaneous presence of two or more large amino acids is not tolerated. It is precisely this constraint which makes for cyclic variation: successive mutants within a subtype are always a step ahead of herd immunity, while the subliminal presence of bridging antigens ensures that the next subtype will arise in the fullness of time and will be senior in terms of the same herd immunity. The number of subtypes will be the same as the number of dependently variable loci (our selection experiments give a minimum of six; serological evidence sets a maximum of seven), and the full cycle covers, on epidemiological evidence, three score years and ten.

Thus is maintained an admirable equilibrium, whereby the span of human life is balanced against an ordered set of antigenic variants. The system is metastable but self-perpetuating as long as the virus keeps within the strictures of its conformation and we keep using last year's vaccines against next year's epidemic. Prospective vaccines (a mixture of senior antigens, representing all subtypes) would set the virus a problem to which, on past performance, it has no answer.

But I am anticipating results to be reported in the volume celebrating André Lwoff's achievements over the next fifty years. Here I note merely that there is room for mutational cycles in the scheme of things, and applaud the editors for acting on this concept. It is fitting to dedicate a volume to André Lwoff on the triumphal completion of a cycle during which all that can be offered in antigenic challenge has been faced and countered, and that without the aid of a hypothesis. But I still wonder at times why the editors chose to honour this particular achievement out of his long list of major ones.

Hard times with André Lwoff's hemoflagellates

S. H. HUTNER

Like a fish unaware of water until high and dry, so was I unwittingly swimming in the Anglo-American culture. As reflected in its biological laboratories, it was predominantly observational, not experimental. There were exceptions, notably in genetics. But on the whole the resolutely reductionist philosophy of Jacques Loeb stirred few ripples. Then I encountered the remorseless Gallic logic of André Lwoff as it was applied to the reductionism he pursued in concert with other European biologists and biochemists. It shocked me into the rut which, after some attempts at escape, I now strive to deepen and widen. Let me explain.

One rainy afternoon in 1936 in Ithaca, James B. Sumner, the enzymologist, was unhappy about his research, which he did when teaching schedules permitted *and* when rain or snow forbade tennis. (In group photos of the faculty, stretching back many years, Sumner was tennis-shoed in about half the photos. One might thus recall Cornell paleo-weather much as one interprets rings in redwood trees or pollen layers in bogs.) I was a graduate student and had grown to like this crusty, irascible, and, I suspect, lonely man. Perhaps he sensed this because he blurted, as if we were peers in experience and wisdom, that his crystalline catalase was exasperating him: was its prosthetic group heme or a nearly identical porphyrin? For if it was heme he could get neither good analytical figures nor sharp spectra because the hydrogen peroxide used in assaying catalase invariably decomposed some of its heme. An impasse!

I rose to the occasion by suggesting that he might identify the heme—if heme it was—by means of the specific growth responses of a flagellate recently alleged by one A. Lwoff (1934) to require heme or protoporphyrin. And, since the creature allegedly responded to cytochrome, it might use the heme in catalase however bound. Sumner duly inspected the article and had me write to Lwoff for a culture, which promptly arrived. I set about isolating enough heme for growth tests. Starting with a bucket of blood obtained at the town abattoir, I carried the preparation of hemin through crystallization from glacial acetic acid (a still-evocative perfume). The response to hemin was a beautifully hyperbolic curve: Lwoff was right. I was about to filter some catalase when word came that Broh-Kahn in Cincinnati, using a heme-requiring *Hemophilus* (a bacterium), had obtained a satisfactory growth response to catalase. The flagellate abruptly joined the ranks of the unemployed.

Much sewage flowed under bridges. When 1943 arrived to shock me with Lwoff's *L'évolution physiologique*, I had been sensitized. At late-afternoon coffee sessions where Professor Hugh Daniel Reed, head of the Zoology Department of Cornell, had held court à la Oliver Wendell Holmes in *The Autocrat of the Breakfast Table*, evolution was the favorite scientific topic, as Reed pridefully was descended from four generations of New England school teachers and traced his scientific ancestry to Louis Agassiz via Burt G. Wilder. Sometimes we discussed what had been gained in metazoan evolution, seldom how these gains had been made. Reed gently, cunningly, advanced a corollary idea for my benefit: that animals had learned marvelous tricks in evolving from algae, and that zoology would remain a miserable, hollow science until these tricks were elucidated. That I was to investigate algae and protozoa toward this end was imprinted on my brain like a behavior pattern on a Lorenz duckling.

To return to *L'évolution physiologique,* my sympathy lay with Lwoff's colleague who had asserted that he could do things a *Chlamydomonas* couldn't, only to meet the cool reply, "Sans doute. Cependant, que serait devenu cet excellent—et par ailleurs éminent—collègue si on l'avait exposé au soleil dans une solution de nitrate de potassium?"

Unconscionable that one should obtain an excellent meal without work-ing, by merely basking Gauguinesquely in the sun! But I had to agree—why not?—let us study these degenerations that somehow must have promoted the evolution of consciousness and thought! And, as para-sites had gone farthest downhill biosynthetically, Lwoff rightly cherished them as materials for studying basic biosynthetic losses. The hemo-flagellates offered the richest, most accessible array of parasites. More-over, some leishmanias and trypanosomes, in plumbing the depths of biochemical depravity, had compounded their felonies by becoming fearsome pathogens. Yet they were highly successful. To rephrase Pangloss, everything was for the worst, biosynthesis-wise, in this most improbable of all thermodynamically possible worlds. My cultivations of photosynthetic bacteria and algal flagellates were no more than exer-cises to develop a technique for the attack on the secret metabolic sins and deadly accomplishments of the hemoflagellates. Lwoff had matter-of-factly accepted their pathogenicity and seemed uninterested in cures —another attitude shocking in an M.D. from l'Institut Pasteur.

I must sketch something of the *mise en scène* to explain why, for an American, this resolve to study hemoflagellates for their own sake was a hard but logical decision. Theobald Smith had indeed reached much the same general conclusions about parasitism; his influence lives on in such workers as William Trager. But at that time such dis-ciples were few. Parasitology was dominated by eminences frozen in a morphological mold, precisely as T. H. Huxley had predicted. They were tirelessly interested in structure and in vagaries of life cycles, and cared not at all how or why parasites did what they did. That is to say, to study parasitism as a biochemically controlled evolutionary process was not a legitimate occupation. The biochemical-genetic revolution, enmeshing ever-broader areas of biology in mechanisms, was more than a decade in the future. True, the hemoflagellates were biochemically degenerate, but, as Theobald Smith had pointed out, it *paid* efficient parasites to be degenerate.

At this time biochemical obscurantism was in disordered retreat after Sumner had won his battle to have crystalline urease recognized as a pure protein. But to study evolution with the tools of chemistry and physics was a pursuit for some vague future. Above all, it was Lwoff and the Dutch proponents of comparative microbiology-biochemistry who proclaimed that biochemical evolution was to be studied now. Hans Zinsser (1940) charmingly described this era from the standpoint of a young physician intent on a research career, bringing out how friendships with French scientists guided much of his development.

Hence Lwoff had me hooked. Some three decades later we may ask how his ideas have stood up and how his bloody pets have fleshed out his theories. He has no need to reproach himself.

First, the hemoflagellates have achieved recognition as indeed being in all likelihood the best materials for studying the evolution of parasitism above (I hastily add!) the viruses. The proliferation of reviews attests to this; they include Hutner, Fromentin, and O'Connell (1968), Kidder (1967), Newton (1968), Rudzinska and Vickerman (1968), and Trager (1968), besides the vast literature on chemotherapy and the pathogens, wherein speculations about biochemical evolution are prominent, as in Hoare (1966), Deane (1969), and Fulton (1969).

Lwoff's taxonomy has withstood the erosion of time—well, not entirely. Thus his generic term *Strigomonas* is being pushed toward the bionomic scrap heap, to be superseded by *Crithidia* (Hoare and Wallace, 1966)—an Anglo-American taxonomic commandment that is being loyally resisted, for example, by Ranque et al. (1968). Some years ago André complained vehemently that as an editor I had countenanced demotion of *Strigomonas*. I shrugged my shoulders and hoisted an eyebrow—wasted Gallic gestures because our colloquy was over the telephone. I smiled: he still loved these hemoflagellates; he too had been imprinted (by Chatton) down to the thalamus, the ancient emotion-controlling center of the brain. His *Strigomonas oncopelti,* adduced as an example of extreme simplicity in nutritional requirements, hence by implication the least biosynthetically degenerate of available hemoflagellates, has mysterious symbionts (Newton and Gutteridge, 1967); its simplicity is illusory.

The heme requirement is becoming a standard demonstration in teaching parasitology (Hutner and Balamuth, 1970). The ability that Lwoff noted of *Crithidia*—all right, *Strigomonas*—to utilize protoporphyrin and hemin interchangeably means that it is endowed with iron chelatase, which inserts the iron into the porphyrin ring; this story, with *Strigomonas* playing a part, is traceable from a review (Burnham, 1969).

Another casualty is the attribution of essentiality to ascorbic acid for cultivating leishmanias and trypanosomes. Korn, von Brand, and Tobie (1969) grew *Trypanosoma cruzi* for a year with stearate. Presumably the ascorbic acid was needed as an antioxidant to avoid rancidity when the fatty acid requirement was satisfied by blood, and perhaps also to keep other nutrients properly reduced.

The peptone requirement concealed a biochemical treasure. *Crithidia fasciculata* (Lwoff's *Strigomonas culicidarum*) and quite probably hemoflagellates in general ("Trypanosomatidae") respond uniquely to biopterin, whose precursor in them is folic acid. In man and the ciliate *Tetrahymena,* biopterin is made from purine; a biosynthetic block then intervenes between biopterin and folate—that is, man can make biopterin. Biopterin (rather, reduced derivatives) is the coenzyme for hydroxylations leading from phenylalanine, tyrosine, and dopa to norepinephrine and epinephrine, also from tryptophan to serotonin. *Crithidia* is thus uniquely valuable (Frank, Baker, and Hutner, 1970) for investigating the neurohormones that so profoundly affect the activities and emotional tone of the central nervous system. Biopterin was isolated simultaneously as one of the pigments of the venerated eye-color ("sepia") mutants of *Drosophila. Crithidia* thus has been much used as a tool in *Drosophila* genetics and for exploring the pigments of butterflies, amphibia, and reptiles (Blakley, 1969; Ziegler, 1965).

It is nicely appropriate that the very organisms in which we have so deep an emotional and intellectual commitment should themselves be means for investigating emotion and intellect. And, as a footnote on unforeseen ways in which the Puritan ethic is confounded, Noguchi may be remembered for the hemoflagellates he isolated, as a pastime, from mosquitoes and milkweed bugs when he vacationed in the Catskills, not for his heroic but misleading work with yellow fever and leptospires.

Lwoff's pioneering on the nutrition of hemoflagellates was the indispensable initial attack on an unexpectedly formidable problem: the full definition of the nutritional requirements of the mammalian trypanosomes. Little progress has been made in resolving the dilemma attending research on the *Trypanosoma brucei* group: cultivated strains lose infectivity; infectious strains are virtually uncultivable. Solving this problem, as Lambrecht (1969) notes, depends on understanding the ecology of the parasite in the fly vector as contrasted with the mammalian host. Most of the mammalian leishmanias and stercorarian trypanosomes are readily cultivable in blood media, and media for *T. cruzi* are almost fully defined. A completely defined medium for a mammalian trypanosome, however, has so far eluded all investigators.

Was Lwoff justified in largely ignoring chemotherapy? Almost certainly. Hemoflagellates remain formidable enemies. African trypanosomiasis threatens to escape control as hints appear of the emergence of drug-resistant strains and of extensive host reservoirs in wild game. Chagas' disease, which devastates much of Brazil and other Latin American countries, defies chemotherapy or immunization, and again the diversity of wild carriers implies that its eradication is centuries away. Mucocutaneous leishmaniasis is a dreadful danger from Central America to Paraguay. Detailed, fundamental knowledge of these adversaries is needed: they will be with us for a long time. The Noguchi-Lwoff *Strigomonas-Crithidia* is being used by at least three major American pharmaceutical houses in screening for antiprotozoal and cytotoxic drugs.

Curiously, too, the pathogens have contributed importantly to molecular biology. Puromycin, originally isolated as an antiprotozoal agent (it proved too toxic), is a standard reagent for blocking protein synthesis, and ethidium (homidium), successful against *Trypanosoma congolense,* is a reagent for fractionating double-stranded DNA. One cannot yet predict from molecular biology whether a particular compound which can bind RNA or DNA, or can interrupt protein or nucleic acid metabolism in some other way, will be a practical drug against tumors or microbes. The existence of the oncogenic viruses may blur future distinctions between an antiviral and an antitumor agent. Rather than philosophize further along these lines, I shall say only that we need to make the most of every investigative tool, and that Lwoff did science a great service when he conceived a theoretical framework for justifying study of the hemoflagellates as exemplars of evolutionary principles.

The cultural debt, I am happy to say, has not been entirely onesided. On a summer visit Lwoff encountered my staff of long-stemmed American beauties, several of them wearing shorts. He demanded photographs. I sent group pictures of them to France with the notation that I hoped they would inaugurate a new era in French laboratory technique. Hélas, this era is still in France's future.

One must, in conclusion, mention another contribution of Lwoff's: *style*. In this connection Hans Zinsser's words about Nicolle are apropos:

Nicolle did relatively few and simple experiments. But every time he did one, it was the result of long hours of intellectual incubation during which all possible variants had been considered and were allowed for in the final tests. Then he went straight to the point, without wasted motion. That was the method of Pasteur, so it has been of all the really great men of our calling, whose simple, conclusive experiments are a joy to those able to appreciate them.

REFERENCES

Blakley, R. L. (1969). *The Biochemistry of Folic Acid and Related Pteridines*, Amsterdam, North-Holland Publishing Co.; New York, John Wiley-Interscience.

Burnham, B. F. (1969). Metabolism of porphyrins and corrinoids, in *Metabolic Pathways*, edited by D. M. Greenberg, 3rd Ed., vol. 3, pp. 403–537, New York, Academic Press.

Deane, M. P. (1969). On the life cycle of trypanosomes of *lewisi* group and their relationships to other mammalian trypanosomes, *Rev. Inst. Med. Trop. São Paulo*, 11, 34–43.

Frank, O., H. Baker, and S. H. Hutner (1970). Antagonists of growth inhibition of *Crithidia* by allopurinol, a guanine analog, *J. Protozool.*, 17, 153–158.

Fulton, J. D. (1969). Metabolism and pathogenic mechanisms of parasitic protozoa, in *Research in Protozoology*, edited by T. T. Chen, vol. 4, pp. 389–504, Oxford, Pergamon.

Hill, G. C., and W. A. Anderson (1970). Recent advances in characterizing the electron transport system in Trypanosomatidae, *Exp. Parasitol.*, in press.

Hoare, C. A. (1968). Evolutionary trends in mammalian trypanosomes, *Adv. Parasitol.*, 5, 47–91.

Hoare, C. A., and F. G. Wallace (1966). Developmental stages of trypanosomatid flagellates: a new terminology. *Nature*, 212, 1385–1386.

Hutner, S. H., and W. Balamuth (1970). In *Experiments and Techniques in Parasitology*, edited by A. J. MacInnes and M. Voge, San Francisco, W. H. Freeman.

Hutner, S. H., H. Fromentin, and K. M. O'Connell (1968). Some biological leads to chemotherapy of blood protista, especially trypanosomatidae, in *Infectious Blood Diseases of Man and Animals*, edited by D. Weinman and M. Ristic, vol. 1: *Diseases Caused by Protista*, pp. 175–216, New York, Academic Press.

Kidder, G. W. (1967). Nitrogen: distribution, nutrition, and metabolism, in *Chemical Zoology*, edited by G. M. Kidder, vol. 1: *Protozoa*, pp. 93–159, New York, Academic Press.

Korn, E. D., T. von Brand, and E. J. Tobie (1969). The sterols of *Trypanosoma cruzi* and *Crithidia fasciculata*, *Comp. Biochem. Physiol.*, 30, 601–610.

Lambrecht, F. L. (1969). Ecological and physiological implications in the selection of trypanosome strains during cyclic transmission, *J. Trop. Med. Hyg.*, 72, 203–209.

Lwoff, A. (1934). Die Bedeutung des Blutfarbstoffes fur du parasitischen Flagellaten, *Zentralbl. Bakt. Parasitol. (I Abt. Orig.)*, 130, 193–200.

Lwoff, A. (1943). *L'évolution physiologique: étude des pertes de fonctions chez les organismes,* Herman et Cie, Paris.

Newton, B. A. (1968). Biochemical peculiarities of trypanosomatid flagellates. *Ann. Rev. Microbiol.,* 22, 109–130.

Newton, B. A., and W. E. Gutteridge (1967). The putative "bacterial endosymbiont" of *Crithidia oncopelti:* a reappraisal, *J. Protozool.,* 14 (Suppl.), 41.

Noguchi, H., and E. B. Tilden (1926). Comparative studies of herpetomonads and leishmanias. I. Cultivation of herpetomonads from insects and plants, *J. Exp. Med.,* 44, 307–326.

Ranque, J., M. Quilici, S. Dunan, and Y. Assadourian (1968). Étude de la structure antigénique de *Strigomonas (Crithidia) fasciculata, J. Protozool.,* 15 (Suppl.), 42.

Rudzinska, M. A., and K. Vickerman (1968). The fine structure, in *Infectious Blood Diseases of Man and Animals,* edited by D. Weinman and M. Ristic, vol. 1, pp. 217–306, New York, Academic Press.

Trager, W. (1968). Cultivation and nutritional requirements, in *Infectious Blood Diseases of Man and Animals,* edited by D. Weinman and M. Ristic, vol. 1, pp. 149–174, New York, Academic Press.

Ziegler, I. (1965). Pterine als Wirkstoffe und Pigmente, *Ergebn. Physiol.,* 56, 1–66.

Zinsser, H. (1940). *As I Remember Him,* Boston, Little, Brown and Co.

Mechanisms for the conservation and efficient use of iodine in man

A. QUERIDO

Three books written by Dr. Lwoff are on my table. The first one, *Recherches biochimiques sur la nutrition des protozoaires,* published in 1932, was written six years before I had the privilege of working in his laboratory. *L'évolution physiologique* appeared a few years after my departure, while the volume of Compton Lectures, with the theme "biological order," dates from 1962. They have in common a deep involvement in problems of evolution. My small project in Dr. Lwoff's laboratory was to develop an assay for nicotinamide, based on the principle of *perte de fonction* by B. proteus of the synthesis of nicotinamide as a constituent for DPN and TPN. This simple project "induced" thinking on general problems in biology and an awareness of the significance of the quantitative approach to molecular processes in biology. A nice example of this is the deterioration of enzyme molecules during metabolic processus.

When through the course of European events in 1940 I had to change my field of interest and became again involved in medicine, the exposure to general problems in biology had profoundly influenced my attitude towards problems of human disease, and led to studies on the regulation of endocrine systems, transfer of signals in the endocrine system, tissue specificity for action of hormones, and molecular disease. It is for this reason that I choose to write a contemplative essay on iodine metabolism, as an example of a highly specialized and regulated biological system, which, as far as we know, has no other purpose than to supply iodine atoms for the manufacture of the thyroid hormones tetraiodothyronine (thyroxine) and triiodothyronine. Both thyroid hormones are of great significance, probably for evolution and for tissue differentiation.

Iodine, in contrast to other trace elements, is in vertebrates localized mainly in one organ, the thyroid. One of the characteristic functions of the thyroid is to concentrate iodide from plasma. The iodine is present in the gland mainly organically bound as an iodoprotein. Being present as an iodoprotein seems unavoidable, because the tyrosine molecules in proteins, under mild conditions *in vitro* at certain concentration of iodine, will bind iodine to mono- and diiodotyrosine. Depending on the relation of tyrosines in protein molecules, the iodotyrosines will couple to diphenyl ethers, the iodothyronines. Under comparable conditions the yield of the hormonally active iodothyronines will, however, be different for different proteins.

There is no evidence that invertebrates are responsive to thyroxine. Turner summarizes the evolution of the thyroid gland as follows: "Iodoproteins have been identified in most invertebrate phyla, but these have a tendency to be localized in hard exoskeletal structures, such as setae, byssus threads, periostracum and in pharyngeal teeth. It is supposed that the source of iodoproteins shifted into pharyngeal region after such materials became of metabolic utility to the organism." Amphioxus has a pharyngeal gland, the endostyle; this is of exocrine nature and contains the iodoprotein, which probably is fragmented in the digestive tract. In the lamprey after metamorphosis the gland is differentiated as a vertebrate thyroid, contains a protease, and is endocrine (see Barrington and Sage, 1963). This is, however, no proof as yet that the iodothyronines of the gland are involved in the metamorphosis of the lamprey, as in amphibians. In teleostei there is more evidence for the significance of thyroid activity for metamorphosis (see Barrington, 1968).

The production and secretion of thyroid hormones is regulated through feedback in a complex system of neurosecretion in the hypothalamus (TRF) and a thyroid-stimulating hormone from the pituitary (TSH). Very recently two laboratories, after many years of difficult isolation, identified TRF as a tripeptide glutamine-histidine-proline active at nanogram level (see Bowers et al., 1969). The thyroid-stimulating hormone is a glycoprotein of molecular weight 25,000 but is not yet sharply identified. The transfer of information and the mechanism of regulation between thyroxine, TSH, and TRF on the molecular level are unknown.

This complex system and its relation to growth and metamorphosis have been studied extensively in amphibians in the last decade. It was shown (see Etkin, 1968) that in the final analysis the metamorphic process is controlled by the hypothalamic neurosecretory apparatus.

The central position of thyroid hormones in metamorphosis, especially in the anurans with their powerful leg action, means that thyroid function was one of the factors that permitted the evolutionary change from water to land. An impressive amount of data has recently been collected on the changes in biochemical processus during metamorphosis. Most processus, tail resorption included, seem to depend on initial protein synthesis manifested by synthesis of enzymes. Tail atrophy was prevented by actinomycin D (see Frieden, 1968). Recently this mechanism of action was also demonstrated for protein synthesis in mitochondria and microsomes isolated from mouse liver.

Many investigators have studied the specificity of the thyroxine molecule and the significance of the (large) iodine atoms. Triiodothyronine, lacking iodine in the 5' position, is in tadpole metamorphosis and in the rat more active than 3,5,3',5'-tetraiodothyronine (thyroxine). Although the concentration of triiodothyronine in human blood is only one tenth of that of thyroxine, its much shorter half-life and higher biological activity make it account probably for 50% of the total thyroid hormone requirements. An interesting synthetic compound is 3'-isopropyl-3,5-diiodothyronine, which maintains the steric configuration of the triiodo compound and is the only compound known that is more active than triiodothyronine, both in metamorphosis and in the rat (Greenberg, Blank, and Pfeiffer, 1963).

For a century it has been known that iodine deficiency affects the development of the human fetus seriously and leads to endemic cretinism. This syndrome is characterized by damage and retardation of the central nervous system, manifested by oligophrenia, deafness, deaf-mutism, and spastic diplegia. There is also abnormal skeletal development. The author has seen areas in the central highlands of western New Guinea and in the Andean Mountains where 10% of the children born was affected (Choufoer, van Rhijn, and Querido, 1965; Querido, 1969). In these areas the daily intake of iodine was 5–10 μg/day, which is one tenth of the minimal supply needed. It is clear that thyroid hormones in mammals also have a highly significant function in the process of development.

The mammalian organism has several mechanisms which provide it with means to be extremely thrifty with iodine. In other words, there is a buffer system between the environment, which offers a varying supply of iodine, and the tissues, which are in daily need of a constant supply of thyroid hormones. We will discuss these highly specialized mechanisms, mostly on the basis of information derived from studies on human beings.

A most important buffer is the presence of a stock of thyroglobulin, the "specialized" iodoprotein. In a human thyroid there generally is about 10 mg of organically bound iodine, while the required daily supply of iodothyronines amounts to an iodine equivalent of 70 μg/day. Of this a variable fraction is recirculated to the thyroid, so that its stock can last for several months. The thyroid is the only endocrine which keeps a supply; all other endocrine glands mainly manufacture to order.

In Fig. 1 the successive steps in the synthesis and secretion of thyroid hormone are presented in a highly schematized form. It is interesting to know that these steps (indicated by small circles) have largely become known through studies on human beings who presented genetically determined thyroid disease. They were typical "experiments of nature," informing us about crucial conditions in thyroid hormone synthesis.

The first step is a fortyfold concentration of iodide through the iodide pump of the thyroid, which apparently brings the cellular concentration high enough for efficient organification. A nonfunctioning iodide pump therefore leads in man to hypothyroidism, as was described by Stanbury and Chapman (1960). The condition has nothing to do with defective thyroxine synthesis as such, because the disease can be overcome by increasing the supply of iodide only. The pump mechanism is also present in the salivary gland, the gastric mucosa, the mammary glands, and the placenta. It supplies the fetus and the newborn with relatively large amounts of iodide. The mechanism is interesting for

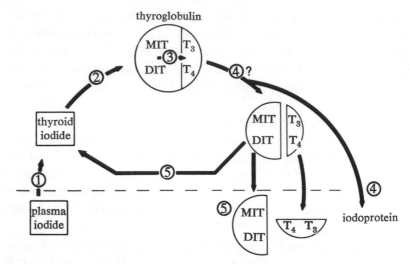

Figure 1. Steps in the formation of triiodothyronine (T3) and tetraiodo-thyronine (thyroxine = T4). 1, iodide pump; 2, organification; 3, coupling of iodotyrosines; 4, proteolysis; 5, deiodination. In the case of steps 4 and 5, what happens when these reactions are absent is also indicated. In step 4 the result is iodoprotein in the serum; in step 5 the iodotyrosines appear in the circulation.

special reasons. Over a range of about 0.05–0.8 µg iodide per 100 ml plasma (dependent on daily iodine intake), the mean daily absolute uptake in the thyroid gland remains constant, probably through auto-regulation of the iodide pump by the gland (see Halmi, 1964; Fisher, 1969). If indeed the extracellular or intracellular iodide concentration regulates iodide transport, there is a biological regulation dependent on an iodine complex which is not thyroxine!

Centrally in the process of thyroid hormone formation stands the thyroglobulin, a glycoprotein having a molecular weight of approximately 650,000, and structured in subunits. The average molecule contains 0.5% iodine; this means 25 iodine atoms per molecule, which has 110 tyrosyl residues (see Robbins and Rall, 1960). The question is whether thyroglobulin is a highly specialized protein of which the tertiary structure favours iodothyronine formation. Data obtained with human low-iodine thyroglobulin (see Crombrugghe et al., 1967) show that *in vitro* iodination yields an iodo-amino acid composition identical with preparations from normal human glands. When thyroglobulin, however, is unfolded, the yield of thyroxine at low and medium levels of iodine is less than that obtained with thyroglobulin in aqueous solution (Salvatore, 1969). These data suggest that no specific enzymes are required in the coupling of iodotyrosines to iodothyronines and that the tertiary structure of thyroglobulin expresses specialization for thyroxine synthesis. For patients with a "coupling defect," it leads to the conclusion that the cause is not the absence of a "coupling enzyme," but rather the presence of an abnormal thyroglobulin.

Much work has been done about the process of iodination as such, which is catalyzed with a peroxidase. There are noteworthy aspects in this reaction, such as the inhibition by an intermediary product (Taurog and Lothrop, 1969), but it is unknown whether this enzyme acts only for the production of iodine in a form which is fit for iodination, or also creates a more favourable iodination yield per mole of thyroglobulin *in vivo* than the noncatalyzed reaction with iodine does *in vitro*. This question cannot be answered because iodide concentration at the spot of reaction is unknown. Genetically determined absence of peroxidase in human beings leads to hypothyroidism.

The steps required for the release of thyroid hormones are the most interesting from the standpoint of economy in iodine metabolism. We will not discuss the required proteolysis of thyroglobulin, which is ascribed by most to a protease in the thyroid, but by some to lysosomal action. In patients this breakdown is sometimes defective, and large fragments of iodinated protein appear in the blood. In the normal breakdown process the thyroglobulin yields mono- and diiodotyrosine, which are biologically inactive, and tri- and tetraiodothyronine. In a normal thyroid gland, for each iodine atom of tetraiodothyronine (thyroxine) stands a yield of eight atoms contained in biologically inactive iodotyrosines. About 70 μg/day of hormonal iodine is needed, and about 560 μg of iodine is liberated daily in the thyroid in nonactive form. The mammalian thyroid is, however, equipped with a dehalogenase (Roche, Michel, and Lissitzky, 1952) which frees iodide and keeps it within the thyroid. The enzyme is also present in other tissues such as liver and kidney.

The consequences of the absence of dehalogenase for the organism have been studied intensively in man. In a family with congenital sporadic cretinism (hypothyroidism pre- and postnatally) the presence of mono- and diiodotyrosine in the blood was demonstrated. The thyroid

gland contained practically no iodine, and the thyroxine level of the
blood was too low (Stanbury et al., 1955). It was demonstrated that
in the thyroids of these patients (see Fig. 2) dehalogenase activity was
absent (see Querido et al., 1956). The conclusion was drawn that the
disease was characterized by impairment of the economy of iodine
metabolism, and that the biological system for thyroxine synthesis was
intact. That this conclusion was correct could be demonstrated by daily
administration of 1 mg of potassium iodide to a young patient, whereby
parameters for the hypothyroid state were restored to normal (Querido
and Choufoer, 1960).

Many investigators consider it possible that there is another safety

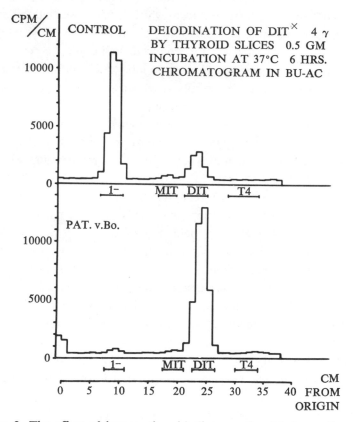

Figure 2. The effect of human thyroid slices on the dehalogenation of diiodotyrosine (DIT). Control: slice of a normally functioning goiter. Lower
figure: slice of a patient with iodotyrosines in the blood.

measure to prevent deficiency of supply of thyroid hormone when the environment offers too little iodine for intake. The idea is that the biologically more active triiodothyronine can be synthesized from mono- and diiodotyrosine, both of which at low iodine supply are available in thyroglobulin, before a level of diiodotyrosine is achieved that favours the formation of tetraiodothyronine. Proof that this mechanism (low serum thyroxine and normal or increased serum triiodothyronine level) actually operates in iodine deficiency is, however, still lacking.

In summary, then, the mammalian organism has a number of powerful mechanisms which protect the individual against irregular or insufficient supply of iodine: a stock of thyroglobulin, an iodide pump adapting to serum iodide levels, and reutilization of most iodine used for the iodination of protein. There are also mechanisms to protect against excessive iodine intake which, however, have not been discussed.

It is clear that these mechanisms are no more or less complicated than other regulatory systems in organisms. Lwoff (1943) made the remark in his book that "l'évolution physiologique est une spécialisation et non un progrès," and "elle semble le résultat d'une évolution progressive; mais son dévéloppement ultérieur la rendra nuisible et sera cause de son extinction." And in a later work (1962) he stated, "The present biological order includes the historical experience of the organism acquired during its evolution. And it is because it includes evolution that it is so highly improbable. Everything unfit has been eliminated."

The fact that these biological mechanisms are, on one hand, so highly improbable and, on the other hand, "probable" because "everything unfit has been eliminated" makes us enjoy studying them. Lwoff's contributions and thoughts on biological order have also been a great stimulus to clinical investigators and have made them recognize the biological perspective of "experiments of nature," while studying mechanisms of disease.

REFERENCES

Barrington, E. J. W. (1968). Metamorphosis in lower chordates, in *Metamorphosis,* edited by William Etkin and Lawrence I. Gilbert, New York, Appleton-Century-Crofts.

Barrington, E. J. W., and M. Sage (1963). *Gen. Comp. Endocrinol.,* 3, 153.

Bowers, C. Y., A. V. Schally, F. Enzmann, J. Boler, and K. Folkers (1969). *Abstract Forty-fifth Meeting of the American Thyroid Association.*

Choufoer, J., M. van Rhijn, and A. Querido (1965). *J. Clin. Endocrinol. and Metab.,* 25, 385.

Crombrugghe, B. de, H. Edelhoch, C. Beckers, and M. de Visscher (1967). *J. Biol. Chem.,* 242, 5681.

Etkin, W. (1968). Hormonal control of amphibian metamorphosis, in *Metamorphosis,* edited by William Etkin and Lawrence I. Gilbert, New York, Appleton-Century-Crofts.

Fisher, D. A. (1969). *Recent Progr. Hormone Res.,* 25, 441.

Frieden, E. (1968). Biochemistry of amphibian metamorphosis, in *Metamorphosis,* edited by William Etkin and Lawrence I. Gilbert, New York, Appleton-Century-Crofts.

Greenberg, C. M., B. Blank, and F. Pfeiffer (1963). *Fed. Prof.,* 22, 621.

Halmi, N. S. (1964). The accumulation and recirculation of iodide by the thyroid, in *The Thyroid Gland,* vol. 1, edited by R. Pitt-Rivers and W. R. Trotter, London, Butterworths.

Lwoff, A. (1943). *L'évolution physiologique,* Paris, Hermann et Cie.

Lwoff, A. (1962). *Biological Order.* M.I.T. Paperback edition, 1965.

Querido, A. (1969). Endemic cretinism: a search for a tenable definition, in *Endemic Goiter,* edited by John B. Stanbury, Pan-American Health Organization, Scientific Publication 193.

Querido, A., and J. C. Choufoer (1960). In *Enzyme Regulations in Clinical Medicine,* edited by A. Gigon, Basil, Benno Schwabe and Cy.

Querido, A., J. B. Stanbury, A. A. H. Kassenaar, and J. W. A. Meyer (1956). *J. Clin. Endocrinol. and Metab.,* 16, 1096.

Robbins, J., and J. E. Rall (1960). *Physiol. Rev.,* 40, 415.

Roche, J., O. Michel, and S. Lissitzky (1952). *Biochim. et Biophys. Acta,* 9, 161.

Salvatore, G. (1969). Personal communication.

Stanbury, J. B., and E. N. Chapman (1960). *Lancet,* 1, 1162.

Stanbury, J. B., A. A. H. Kassenaar, J. W. A. Meyer, and J. Terpstra (1955). *J. Clin. Endocrinol. and Metab.,* 15, 1216.

Taurog, A., and M. Lothrop (1969). *Abstract Forty-Fifth Meeting of the American Thyroid Association.*

Lysosomes and fertilization

E. F. HARTREE

LYSOSOMES

The discovery of lysosomes stemmed from observations made by de Duve and his co-workers on the acid phosphatase in rat liver. When the tissue was gently homogenized in a balanced sucrose solution, it exhibited a low level of enzyme activity; but when it had been stored overnight in the frozen state, the activity was considerably higher.

The existence of lysosomes as distinct cellular entities was established by Appelmans, Wattiaux, and de Duve (1955) and by de Duve, Pressman, Gianetto, Wattiaux, and Appelmans (1955). When the mitochondrial fraction of rat liver was further fractionated, most of the cytochrome oxidase appeared in the denser particles, while acid phosphatase, together with acid protease, RNAase, DNAase and β-glucuronidase, was concentrated in the less dense particles. All enzyme activities were increased when the particles had been frozen. From these experiments emerged the concept of a lysosome: a particle bounded by a lipoprotein membrane which is to a great extent impermeable to intralysosomal hydrolytic enzymes and to external substrates for these enzymes. A feature of lysosomal enzymes is their latency: measured activities are low when assayed under conditions which preserve lysosomal integrity, that is, in neutral iso-osmotic media, at a low temperature, and with short incubation periods. Release of latent enzymes takes place as a result of treatments which disrupt lipoprotein membranes: vigorous homogenization, hypo- and hyperosmotic media, phospholipase A, proteases, low pH, surfactants, organic solvents, and ultraviolet and ionizing radiations.

Five reviews (I–V) covering the development of the lysosome concept are listed in the References at the end of this paper. de Duve's contribution to I covers the early developments, while II, which reports the first international symposium on the subject, demonstrates how rapidly the subject developed. The state of knowledge in 1969 is surveyed in V.

The characterization of liver lysosomes in enzymatic terms implied the existence of hydrolase-carrying organelles. They were identified in liver fractions by Novikoff, Beaufay, and de Duve (1956) as a fairly uniform population of granules which resembled pericanalicular dense bodies (Novikoff, 1959). Methods for locating acid phosphatase histochemically were already available, and granules that stain for this enzyme were soon shown to be very widely distributed among eukaryotic cells. Thus acid phosphatase has become the traditional marker for lysosomes. These observations stimulated the development of histochemical methods for detecting other lysosomal enzymes. The number of known lysosome-associated enzymes is now large (Barrett, 1969; Tappel, 1969). They include DNAase and RNAase, proteases, enzymes that degrade polysaccharides and simpler sugars (e.g., hyaluronidase, sulfatases, glycosidases, and β-glucuronidase), and lipolytic enzymes such as phospholipase A and esterases.

The early views on the formation of lysosomes were as follows. Biosynthesis of hydrolases proceeds in the endoplasmic reticulum and Golgi apparatus. Small vesicles budding off from the latter become the primary lysosomes. These bodies may fuse with phagocytic vacuoles containing endocytosed extracellular material or may themselves engulf cytoplasmic constituents. In these secondary lysosomes digestion occurs, and nondigested residues are either stored in "residual bodies" or released from the cell by exocytosis. According to later views (Novikoff, Essner, and Quintana, 1964), certain types of primary lysosome bud off from the smooth endoplasmic reticulum, while secondary lysosomes (digestive vacuoles) arise when portions of cytoplasm become enclosed within a membrane formed from the rough endoplasmic reticulum.

The functions of lysosomes are summarized in Table 1. These func-

TABLE 1. FUNCTIONS OF LYSOSOMES

Active aggression against microorganisms, viruses, toxic particles, and macromolecules by leucocytes, macrophages, and reticuloendothelial cells.

Heterotrophic nutrition by intracellular digestion and in certain circumstances by extracellular digestion.

Nutrition under adverse conditions through self-digestion (autophagy); intracellular scavenging; self-clearance of dead cells; clearance of dead cells by heterophagy.

Readsorption in kidney and bladder; processing of secretory products in gland cells (thyroid); resorption of bone and connective tissue fibres.

Developmental processes: metamorphosis, regression, involution, fertilization, and embryonic development.

tions are essentially digestive, and they depend upon the ability of the lysosome membrane to fuse with other membranes. de Duve (1969) emphasizes that the lysosome should be regarded not as a body but as part of a vacuolar digestive system in which so far undetectable transient fusions and fissions may be occurring. Such a system assures the protection of the cellular structure from the damaging effects of lysosomal hydrolases.

The main physiological functions of lysosomes come under the headings of defence against foreign materials, nutrition, scavenging, enzyme secretion, and tissue remodelling (Table 1). In metamorphosis, tissue regression, and involution there is evidence not only for the programmed release of lysosomal enzymes but also for striking growths of lysosome populations in anticipation of requirements for digestive enzymes. The importance of lysosomes in pathology arises from their indiscriminate appetite for foreign materials together with the fact that such materials may labilize the lysosome membrane. Intracellular release of lysosomal enzymes is a feature of tissue damage caused by hypoxia, hyperoxia, thermal injuries, ischaemia, bacterial endotoxins, and ultraviolet and ionizing radiations (Ref. IV). The uptake by lysosomes of many compounds, including drugs and dyes, can be demonstrated by fluorescence microscopy. Thus the brilliant orange-red fluorescence of lysosomes in cells vitally stained with acridine orange has had wide application, as has also their positive periodic acid-Schiff reaction. Lysosomes containing fluorescent markers may be selectively damaged in intact cells by light of appropriate wavelengths, and in this way the damaging effects of enzyme release can be studied (Allison and Young, 1969). Antihistamines, corticosteroids, and antimalarials such as chloroquine stabilize lysosomal membranes, and their therapeutic properties are in some cases consistent with a protection of cells from the adverse effects of "leaking" hydrolases.

The lysosome was originally pictured as a bag containing a solution of hydrolytic enzymes. Thus damage to the lysosome membrane would cause simultaneous release of these enzymes in soluble form. It later became apparent that lysosomes have a granular matrix, that damage to the membrane causes unequal degrees of "release" of hydrolases, and that released hydrolases are to some degree still attached to membranous material (Tappel, 1969). There is now evidence that glycolipid groups are important structural features of lysosomes. They can bind certain foreign substances and ions that are taken up by lysosomes; they may also serve to bind hydrolases in such a manner as to render them comparatively inert (Barrett and Dingle, 1967; Koenig, 1969; Shamberger, 1969).

FERTILIZATION

Monroy introduces his book *Physiology and Biochemistry of Reproduction* as follows:

In his classical work, *The Cell in Development and Inheritance*, E. B. Wilson wrote: "The essential phenomenon of fertilization is the union of a sperm nucleus, of paternal origin, with an egg nucleus, of maternal origin, to form the primary nucleus of the embryo." This definition indeed outlines the basic feature of fertilization, namely that of bringing together the genetic information contained in the nuclei of the two gametes. This very fact makes of fertilization not only the beginning of morphogenesis but also one of the fundamental instruments of evolution.

Nuclear union, however, is the culmination of a series of sperm-egg interactions that begins before the spermatozoon makes contact with the egg. For the purposes of this review I define fertilization as the following sequence of events: (i) interaction at a distance between gametes, (ii) attachment of the sperm cell to the egg envelope followed by penetration through that envelope, (iii) morphological and metabolic changes in the egg that are a response to the penetrating spermatozoon, and (iv) mitosis and cell cleavage. The first two events may involve the release from eggs of substances that stimulate sperm motility or may, in ways not yet understood, bring the sperm cell to a condition which enables it to penetrate the egg investments. Event iii is described as activation: it may also be induced parthenogenetically, especially with invertebrate eggs, by a wide range of agents. Such experiments have revealed that the egg contains in latent form much of the equipment for completion of stage iv and for subsequent embryonic development.

MALE AND FEMALE GERM CELLS

Spermatozoa vary widely in size, and the heads also vary considerably in shape (Fawcett, 1970). A typical mammalian spermatozoon is about 50 μ long with a flattened oval head approximately 5 $\mu \times$ 2.5 μ (Fig. 1, 1). In addition to the head (nucleus) and tail, there is usually a wider tail-section adjacent to the head: the midpiece (Fig. 2a). The anterior two thirds of the head is covered by a thin cap, the acrosome, while the posterior part is covered by the even thinner postnuclear cap. The acrosome consists of an acrosomal matrix which is bounded by inner and outer acrosomal membranes. The entire sperm cell is surrounded by a very thin plasma membrane. The posterior portion of the acrosome, the equatorial segment, appears as a roughly triangular region in Giemsa-stained spermatozoa (Fig. 1, 1) or in unstained spermatozoa under phase-contrast. During spermateliosis, the emergence of a spermatozoon from a spermatid, part of the spermatid cytoplasm is retained, as the cytoplasmic droplet, at the neck. As the spermatozoon moves through the epididymis the droplet migrates to the distal end of the midpiece and, after ejaculation, is normally free in the seminal plasma (Bloom and Nicander, 1961; Dott and Dingle, 1968).

The egg of a placental mammal is enclosed in a thick translucent membrane, the zona pellucida. The newly ovulated egg is embedded in a mass of granulosa (follicular) cells (Fig. 2b, Fig. 1, 3). Of these the outer layer, the cumulus oophorus, is held together rather loosely within a matrix of hyaluronic acid. The inner corona radiata is a more compact

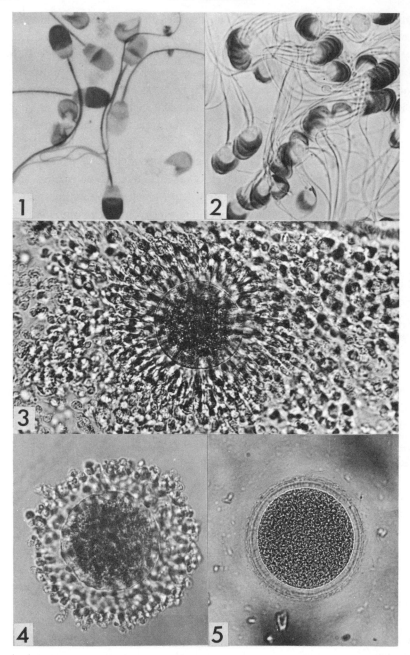

Figure 1

arrangement of cells. Eggs of nonmammals have no zona membrane, but they may be surrounded by jelly (echinoderms, some amphibia) or encased in a hard membrane, the chorion (fish).

In attempting to summarize possible relationships between lysosomes and fertilization I am presenting a very incomplete picture of the latter. For this reason I list reviews VI–XI.

LYSINS IN ACROSOMES

In toads and some invertebrates the acrosome appears to originate from the Golgi apparatus as a proacrosomal granule in the early spermatid. As the spermatid develops, the granule becomes attached to the nuclear membrane, where it undergoes changes and finally becomes the acrosome of the mature spermatozoon (Dan, 1967). Examination of squashes of guinea-pig testis vitally stained with acridine orange shows that early spermatids contain numerous red-fluorescent granules which coalesce during maturation to form two aggregates (Allison and Hartree, 1970). One becomes the prominent proacrosome of the late spermatid; the other moves to the opposite pole of the nucleus and may become associated with the cytoplasmic droplet. The acrosomes of mature guinea-pig and rat spermatozoa fluoresce orange-red when vitally stained with acridine orange (Bishop and Smiles, 1957; Allison and Young, 1964).

Because the acrosome originates in the Golgi zone Bowen (1924) suggested that it might carry secretory enzymes into the egg. Parat (1933) wrote, "La cellule sexuelle mâle se comportant en somme comme si elle élaborait un énorme grain de ferment, l'acrosome, qui vient coiffer son extrémité céphalique." Subsequently, elegant studies on invertebrate fertilization by Colwin and Colwin (1967) and Dan (1967) revealed, besides great varieties of acrosomal morphology, some interesting "ac-

---

←Figure 1. *1*. Ram spermatozoa treated with Hyamine 2389 (Giemsa staining). Sperm heads without acrosomes stain only at the equatorial segment. PAS staining gives similar results (720×). *2*. Guinea-pig spermatozoa stained for chlorobromoindoxyl acetate esterase. The stain is limited to the acrosomes (585×). *3*. Freshly ovulated rabbit egg within granulosa cell mass (230×). *4*. Egg, as in *3*, treated with hyaluronidase to remove cumulus oophorus (150×). *5*. Egg, as in *3*, treated with LGP from bull spermatozoa. All granulosa cells have been removed, and the zona pellucida has been attacked (120×).

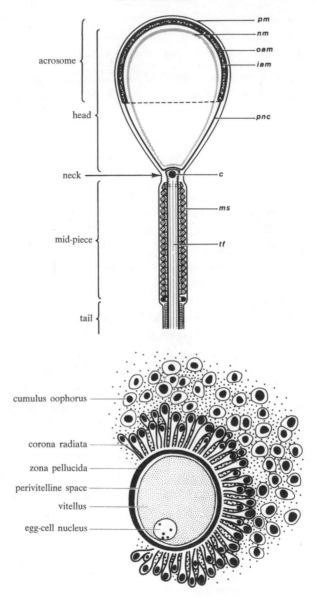

Figure 2. Diagrammatic representations of (*a*) head and midpiece of a mammalian spermatozoon, and (*b*) freshly ovulated mammalian egg with granulosa cell investments. *pm*, plasma membrane; *oam*, outer acrosomal membrane; *iam*, inner acrosomal membrane; *nm*, nuclear membrane; *pnc*, postnuclear cap; *c*, centriole; *ms*, mitochondrial spiral; *tf*, tail fibres.

rosome reactions." These are of two types. In one the acrosomal membrane fuses with the egg membrane and passage of the sperm nucleus into the egg is facilitated by lysins released from the acrosome. The second type involves extrusion of a fibrous, often rod-like, organelle from the anterior end of the sperm head. This structure appears to carry lysins to the egg membrane. Wada, Collier, and Dan (1956) observed that a lysin could be isolated from sea water in which *Mytilus* spermatozoa had been stored provided that the acrosome reaction (rod extrusion) had taken place. If extrusion had not occurred, the lysin was retained in the spermatozoa.

Of the granulosa cells that surround newly ovulated rabbit eggs the cumulus is dispersed *in vitro* by a concentrated suspension of rabbit spermatozoa (Pincus, 1930; Yamane, 1930). McLean and Rowlands (1942) showed that this was due to release of hyaluronidase from the sperm cells. Austin (1960) observed that digitonin simultaneously caused detachment of acrosomes and release of hyaluronidase; subsequently Mancini et al. (1964) established, by immunofluorescence techniques, that in bull spermatozoa hyaluronidase is localized in the acrosomes. However, pure hyaluronidase has no effect upon the corona radiata or on the zona pellucida. Austin and Bishop (1958) noted a loosening or detachment of rodent acrosomes when the spermatozoa were close to the zona pellucida. They postulated the presence of a "zona lysin" in the perforatorium, an apical body, or space, situated between the acrosome and the anterior region of the sperm nucleus. Attempts to isolate such a lysin were, however, unsuccessful (Austin, Ref. VIII).

INVESTIGATIONS ON MAMMALIAN ACROSOMES

Acrosomes give a positive periodic acid-Schiff reaction for carbohydrate (Wislocki, 1949). Clermont, Glegg, and Leblond (1955) treated guinea-pig spermatozoa with dilute alkali and found that the PAS-reactive material dissolved. After acid hydrolysis the extract contained galactose, mannose, fucose, and hexosamine.

Composition of Ram Sperm Acrosomes. R. R. Hathaway and I compared the effects of detergents and alkali as means for removing acrosomes from bull and ram spermatozoa (Hathaway and Hartree, 1963). Figure 3 illustrates the effects of graded concentrations of alkali

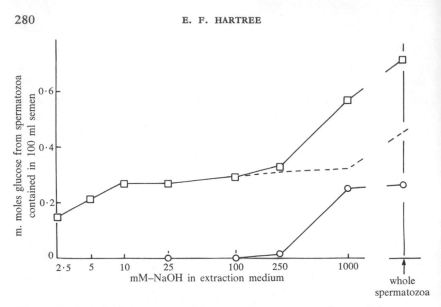

Figure 3. Carbohydrate extracted from ram spermatozoa by graded concentrations of alkali. □, Orcinol-reactive carbohydrate as glucose. ○, Bound deoxyribose converted to its glucose equivalent in the orcinol reaction. The pecked line is a difference curve (Hartree and Srivastava, 1965).

on the extent of solubilization of ram-sperm carbohydrate. Three types of carbohydrate are present. The plateau between 0.01 and 0.25 N sodium hydroxide corresponds to dissolution of the acrosome. Gross morphological changes occur only at higher concentrations of alkali, which dissolve nucleic acid (deoxyribose) but leave undissolved a fine granular precipitate containing carbohydrate that is, after hydrolysis, predominantly glucose (Table 2). Membranous material which was strongly PAS positive was also removed from ram spermatozoa by treatment with cetyltrimethylammonium bromide, but there was obvious erosion of other parts of the sperm cell and also separation of heads from tails.

    In a search for methods for detaching morphologically recognizable acrosomes from ram spermatozoa with minimal damage to the cells, P. N. Srivastava tested a range of surfactants. The most promising results were obtained with the Rohm and Haas product Hyamine 2389, a cationic detergent (Hartree and Srivastava, 1965) (Fig. 1, 1). Subsequent centrifugal fractionation led to separation of acrosomes from the

TABLE 2. SUGARS IN ACID HYDROLYSATES OF FRACTIONS FROM WASHED RAM SPERMATOZOA

Sugar	Acrosomal preparation (Hyamine 2389) A	Material soluble in 0.0125 N NaOH B	Material soluble in 1 N NaOH	Material insoluble in 1 N NaOH
Glucose	n.d.*	n.d.	−	+++
Fucose	(1.0)	0.9	−	−
Galactose	3.7	3.9	−	+
Mannose	6.3	5.9	−	+
Galactosamine	2.8	2.8	−	−
Glucosamine	6.2	6.2	−	−
Sialic acid	1.9	0.9	−	−
Deoxyribose	n.d.	n.d.	+	−

* Not detected.
Analyses are expressed as molar proportions with reference to fucose in A.

heavier spermatozoa. These detached acrosomes were unstable, and on dialysing them an almost clear solution (A) was obtained. The composition of this extract was compared with that of acrosomal material extracted from ram spermatozoa by 0.0125 N sodium hydroxide (B). Lipid analyses on A and B gave comparable results, and after hydrolysis the contents of monosaccharides (Table 2) and amino acids were very similar. From this we deduced that the analytical figures were probably representative of the intact acrosome and that LGP-A in fact contained essentially only acrosomal substance (Hartree and Srivastava, 1965).

After paper electrophoresis of A and B we found that almost all the protein, lipid, and carbohydrate was, in each case, present in a single diffuse band. This lipoglycoprotein (LGP) is thus a major component of ram acrosomes. Miller and Mayer (1960) came to a similar conclusion after analysing alkaline extracts of bull spermatozoa.

Lytic Activity of Extracts of Acrosomes. The localization of hyaluronidase in acrosomes, the possible association of acrosomes with Austin's zona lysin, and indeed the obvious function of acrosomes in the fusion of gametes of lower animal forms suggested that LGP-A might contain egg lysins. With the cooperation of Dr C. E. Adams we tested the hypothesis with rabbit eggs. In this species spermatozoa penetrate the zona pellucida while the egg is still closely invested by granulosa cells (Fig. 1, *3*). When eggs are placed in hyaluronidase, only the cumulus cells are dispersed (4). However, contact with LGP-A caused dissolution of cumulus and corona cells (5), followed by lytic attack on the zona pellucida and even fragmentation of the vitellus (Srivastava, Adams, and Hartree, 1965). The LGP-A from ram, bull, and rabbit spermatozoa had comparable lytic actions on rabbit eggs. These preparations showed protease activity, which, with the presence of hyaluronidase, would explain their effects upon eggs.

In vivo, a single rabbit spermatozoon penetrates between granulosa cells without causing a noticeable dispersion. Thus any sperm lysins exert a highly localized action. Similarly the fissure in the zona pellucida by which a sperm cell reaches the perivitelline space has a cross section that is only marginally larger than that of the sperm head (Bedford, 1968). In our experiments eggs were exposed to a higher concentration of acrosomal factors than would be the case *in vivo,* and the more drastic effects are not surprising. An intermediate situation is the comparatively widespread disintegration of egg membranes that may be induced by lysins from a single invertebrate spermatozoon (Colwin and Colwin, 1960).

Acrosomes as Lysosomes. After the experiments just described were published, Dr A. C. Allison suggested to me that there were now reasons for regarding the acrosome as a lysosome: (1) hyaluronidase and proteases are enzymes that occur in lysosomes (Aronson and Davidson, 1965; Boumer and Gruber, 1964; Barrett, 1967), and (2) a PAS reaction and a red fluorescence after vital staining with acridine orange are characteristic properties of lysosomes. De Duve (1963) had suggested that a necessary step in fertilization could be a release from the spermatozoon of lysosomal-type enzymes. Although at that time no lysosomes had been identified in spermatozoa, the occurrence of acid phosphatase in guinea-pig acrosomes had been established histochemically by Ruffili (1960). Subsequently this enzyme was detected in acrosomes of other species (see Allison and Hartree, 1970).

Dr Allison and I looked for five lysosomal enzymes in LGP-A: acid phosphatase, aryl sulfatase, β-N-acetylglucosaminidase, phospholipase A, and proteases. All were present. For these experiments we devised a procedure for washing ram spermatozoa which removed virtually all the cytoplasmic droplets. This step was obligatory because Dott and Dingle (1968) had discovered that these droplets also carry lysosomal enzymes. By histochemical methods we showed that chloro-bromoindoxyl acetate esterase, another lysosomal enzyme (Holt, 1963), is restricted, in guinea-pig spermatozoa, to the acrosomes (Fig. 1, 2). We have been unable to demonstrate acrosomal staining of ram spermatozoa for acid phosphatase although suspensions of broken acrosomes stain by both the Gomori (1952) and the Burstone (1961) methods. Dott and Dingle (1968) report that cytoplasmic droplets are similarly acid phosphatase positive only when disrupted. Unlike rodent acrosomes, the acrosomes of ram, bull, boar, and human spermatozoa fluoresce only weakly with acridine orange.

On the basis of these results we proposed that the acrosome is a specialized form of lysosome which has evolved to facilitate fertilization in multicellular organisms (Allison and Hartree, 1969a, 1969b, 1970). The lysosomal-type enzymes that we found in LGP-A are present also in the seminal plasma. However, our results indicated that with the possible exception of aryl sulfatase the hydrolases in LGP-A preparations represent genuine acrosomal constituents (Allison and Hartree, 1970). Bernstein and Teichman (1968) treated bull spermatozoa with barbituric acid, which caused dislodgement and dissolution of acrosomes and also release of protease and acid phosphatase. Rabbit spermatozoa behaved similarly except that most of the acid phosphatase remained attached to the sperm heads. However, it is uncertain whether barbituric acid caused removal of all acrosomal membranes.

VESICULATION OF ACROSOMES: DIGESTION OF THE ZONA PELLUCIDA

Suggestive evidence that acrosomes play a role in the penetration of rabbit eggs by spermatozoa is provided by electron microscopic studies on changes in the acrosome that precede fertilization *in vivo* (Barros et al., 1967; Bedford, 1968). This acrosomal reaction occurs as a spermatozoon traverses the cumulus and corona cells and is completed when the sperm head reaches the zona. The interpretation of these changes (Barros et al., 1967) is that the plasma membrane and the outer acrosomal membrane undergo multiple point fusions with the result that in sections for electron micrography the entire acrosome, except for the equatorial segment, appears as a chain of vesicles. The same phenomenon was described for rat spermatozoa by Pikó (1967). The consequence of this change is exposure to the environment of both

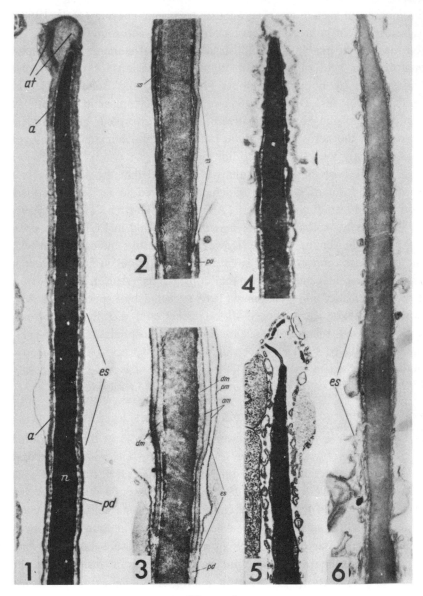

Figure 4.

inner and outer acrosomal membranes. The acrosomal membranes and plasma membrane of a normal ram spermatozoon are depicted in Figure 4, *1*. The close investment of the nucleus by the posterior section of the acrosome, to form the equatorial segment, is well defined.

Vesiculation was first described by Saacke and Almquist (1964) in bull spermatozoa (*2*). In this case vesiculation was unrelated to fertilization since the semen had been collected in an artificial vagina. In the rabbit vesiculation takes place only as a spermatozoon passes among the granulosa cells (*3*). When acrosomes are dislodged from ram spermatozoa with Hyamine, the inner acrosomal membrane remains attached to the head but the equatorial segment is resistant to Hyamine (Fig. 1, *1;* Fig. 4, *4*). This means that the disposition of membranes around the rabbit sperm nucleus after vesiculation has occurred is similar to that around the nucleus of a Hyamine-treated ram spermatozoon. One difference is that in the former case the plasma membrane apposed to the equatorial segment remains in situ and fuses anteriorly with the equatorial segment (Bedford, 1970). Recent electron micrographs suggest that when acrosomes are released from Hyamine-treated ram spermatozoa the acrosomal matrix is sufficiently rigid to retain its shape despite the absence of an inner acrosomal membrane (Fig. 4, *4*). Subsequently the detached acrosomes break down to small vesicles and finally to the soluble LGP (Hartree and Brown, 1970). It has been suggested (Bedford, 1968; Allison and Hartree, 1970) that vesiculation may be a mechanism for ensuring the sequential availability of acrosomal enzymes. Vesicle formation allows the acrosome contents to be released as the rabbit spermatozoon penetrates between granulosa cells, that is, at a time when the enzyme hyaluronidase could play a role. After

---

Figure 4. Electron micrographs of sagittal sections of heads of mammalian spermatozoa. *1*. Ram spermatozoon (21500×). *2*. Ram spermatozoon: equatorial region (36000×). *3*. Guinea-pig spermatozoon: equatorial region. Plasma membrane and acrosomal membranes are clearly defined (41500×). *4*. Bull spermatozoon with vesiculated acrosome (17000×). *5*. Rabbit spermatozoon (vesiculated acrosome) between corona cells (15500×). *6*. Ram spermatozoon treated with Hyamine 2389. Both the inner acrosomal membrane and the equatorial segment remain attached but are damaged (23500×). *a*, acrosome; *am*, inner and outer acrosomal membranes; *at*, acrosomal thickening (apical ridge); *dm*, dense material; *es*, equatorial segment; *pd*, postnuclear dense lamina (postnuclear cap); *pm*, plasma membrane.

vesiculation is complete, presumably only the lysosomal enzymes that remain attached to the inner acrosomal membrane can facilitate passage of the spermatozoon through the zona pellucida.

The zona pellucida of rabbit eggs is digested by proteolytic enzymes and by crude hyaluronidase but not by pure hyaluronidase (Bedford, 1968). Stambaugh and Buckley (1969) suggested that the hyaluronidase and trypsin-like activities in Hyamine extracts of rabbit spermatozoa reside within a single molecular species. This would represent a crude hyaluronidase, but evidence that this enzyme has any effect upon the zona is lacking except for an interesting observation by Konechny (1959) on the oocytes of sectioned cat ovaries. He found that metachromasia of the zona in preparations stained with toluidine blue was abolished in sections that had been treated with hyaluronidase. It is thus possible that although this enzyme does not visibly modify the zona membrane it may induce changes in zona polysaccharides that facilitate proteolysis. When rabbit and monkey spermatozoa are treated with a soya-bean trypsin inhibitor carrying a fluorescent label, the inhibitor becomes attached to the inner acrosomal membrane (Stambaugh and Buckley, 1970). This result is consistent with observations that, unlike other ram sperm proteases (Allison and Hartree, 1970), the trypsin-like enzyme of ram spermatozoa remains firmly attached to the cells after acrosomes have been dislodged by Hyamine (Hartree and Brown, 1970). The trypsin-like enzyme is an obvious candidate for the role of zona lysin: it is however not a typical lysosomal enzyme, being absent from rat-liver lysosomes (Hartree and Brown, 1970).

The equatorial segment, which remains firmly attached to a rabbit spermatozoon when it has reached the perivitelline space, appears to carry hydrolytic enzymes that assist the sperm nucleus to fuse with the egg cortex (Bedford, 1970).

CAPACITATION OF SPERMATOZOA

This phenomenon, discovered by Chang (1951) and Austin (1951, 1952), may be defined as a change in ejaculated spermatozoa that fits them for fertilization. The change occurs when spermatozoa remain in the uterus or Fallopian tubes for several hours. Only capacitated rabbit spermatozoa can fertilize eggs *in vitro*. The phenomenon, which has also been observed in other species, has recently been the subject of reviews by Austin, Chang, and others (see Raspé, 1969).

Although capacitation causes no visible changes in spermatozoa, it is generally believed to involve removal of a surface coating and a possible increase in sperm membrane permeability. The only clear experimental evidence for such changes are reports that capacitation of rabbit spermatozoa results in changes in distribution of charged groups on the plasma membrane surface (Bedford, 1970). Only capacitated rabbit spermatozoa undergo vesiculation. Smith (1951) obtained capacitation of rabbit spermatozoa that had been incubated with fragments of tubal mucosa, while Hamner and Sojka (1968) found that rabbit spermatozoa contained in a Millipore tube filter remained uncapacitated when placed in a rabbit uterus. On the other hand capacitation of mouse, hamster, and cow eggs has been effected in vitro by incubating the eggs in media containing fluid from the fallopian tubes (Bedford, 1970).

In their experiments on the human uterus Garcia-Bunuel and Brandes (1966) noted that just before ovulation many macrophages accumulate between epithelial cells and that they subsequently pass into the lumen. These authors suggested that digestive activities of the macrophage granules provide metabolites for spermatozoa. An alternative possibility is that macrophage hydrolases bring about changes in sperm membranes or in surface coatings that are the basis of capacitation. This idea receives support from the discovery by Ericsson (1969) that rabbit spermatozoa can be capacitated by mule eosinophils. Removal of the zona pellucida from rat eggs allows them to be fertilized by uncapacitated spermatozoa (Toyoda and Chang, 1968). This observation is consistent with the view that lysins essential for penetration of the zona become available only after capacitation, that is, after changes which would allow vesiculation to occur. The nature of the trigger to vesiculation, which in rabbits presumably originates in the granulosa cells, is unknown.

INITIAL REACTIONS OF EGGS TO SPERM PENETRATION

For reasons of ready availability the eggs of echinoderms, fish, and amphibia have been used extensively for studying the development of fertilized eggs.

The outer membrane in these organisms is usually the thin vitelline membrane. Beneath this, and closely apposed to it, is the egg plasma membrane. Many fish eggs have an additional outer membrane, the chorion. Embedded in the egg cortex and close to the inner surface of the plasma membrane are numerous vesicles, the cortical granules (or cortical alveoli). The first visible reaction of echinoderm eggs to sperm penetration is elevation of the vitelline membrane (which now becomes the fertilization membrane) and formation of the perivitelline space. This elevation follows bursting of the cortical granules and discharge of their contents into the perivitelline space. The first obvious physiological change is the dramatic block to polyspermy. In eggs without a zona the block may reside in either the vitelline membrane or the plasma membrane. It is established very rapidly: within about 1 minute in sea-urchin eggs. Among mammalian eggs the block may reside in the zona or at the egg surface. Subsequent changes in fertilized sea-urchin eggs are (1) increases in permeability to water and ions; (2) changes in membrane potential; (3) activation of a membrane-bound ATPase; (4) a marked increase in respiratory activity, which is associated with a swelling of mitochondria, an increase in the ADP/ATP ratio, and an activation of cytochrome oxidase; (5) breakdown of polysaccharides and stimulation of glycolysis; and (6) initiation of protein synthesis (Monroy, Ref. X).

The work of Yamamoto (1961, 1962) on the eggs of fresh-water fish, notably *Oryzias latipas,* sheds light on the nature of the cortical reaction. In eggs of *Oryzias* the cortical granules are 10–40 μ in diameter. Penetration of a spermatozoon initiates breakdown of these granules at the animal pole, the one point in the chorion surface where sperm entry is possible. A wave of breakdown then spreads uniformly in all directions over the cortical surface. Simultaneously the perivitelline space is formed through cortical shrinkage. Yamamoto isolated *Oryzias* cortical granules and characterized them as vesicles containing colloid and bounded by a single lipoprotein membrane.

To account for the cortical reaction one could suppose that a factor carried by the spermatozoon causes lysis of cortical granules near the animal pole. It may then be supposed that breakdown of a granule releases a lytic agent which in turn degrades neighbouring granules. In other words the cortical granules are lysosomes. This mechanism is untenable, however, because when cortical granules are sedimented into the vegetal half (by centrifuging) the *Oryzias* egg can still be fertilized or parthenogenetically activated (see below). Thus, although the spermatozoon enters at the animal pole, cortical granules in the vegetal half break down. In the interstitial protoplasm of the cortex Yamamoto observed "a-granules" which are red and 0.3–0.5 μ in diameter. In over-ripe *Oryzias* eggs a proportion of cortical granules may break down spontaneously, and in the regions of breakdown a-granules are no longer visible. An entering spermatozoon also induces changes in neighbouring a-granules: they acquire a more diffuse outline, enlarge, and disappear. Immediately after these changes cortical granules become slightly pink and discharge their contents into the perivitelline space.

Lysosomes and Membrane Fusion: Effects of Ferrous Ions. Yamamoto observed that low concentrations of ferrous sulfate (10 μM) greatly increased the fragility of isolated cortical granules, while Daniel and Millward (1969) found that similar concentrations of ferrous salt were essential for cleavage of fertilized rabbit eggs *in vitro*. In both cases ferric salts were ineffective. Suggestions that these phenomena have an enzymatic basis come from work by Paigen (1958) and by Revel and Racker (1963) on mammalian phosphoprotein phosphatase and from experiments by Ono and Nojima (1969) on a membrane-bound bacterial phospholipase A. In each case the enzyme appears to exist as an inactive Fe^{3+} complex which becomes active when the iron is reduced. The cortical reaction and cell division involve the restructuring of lipoprotein membranes, a process in which lysolecithin, a product of phospholipase A activity, is believed to be involved (Lucy, 1969). Support for this belief comes from Howell and Lucy's (1969) observation that lysolecithin induces syncytia formation in suspensions of hen erythrocytes. The fact that phospholipase A is a lysosomal enzyme (Blaschko et al., 1967; Reinauer et al., 1968) supports the view that lysosomes (a-granules?) are involved in the cortical reaction and also points to a possible role for lysosomes in egg cleavage. Recently developed methods for isolating sea-urchin cortical granules (Epel, Weaver, Muchmore and Schimke, 1969) could throw light on the role of these particles.

Activation of Eggs: Parthenogenetic Development. In considering this phenomenon it is appropriate to begin with the classical work of

Loeb (1913). He found that slightly hypertonic sea water did not cause the fertilization membrane of sea-urchin eggs to rise, though a gelatinous layer appeared around the egg. Development began, but cell cleavage was defective and the blastula stage was seldom reached. A wide range of agents (activating factors) caused membrane elevation, though the eggs subsequently cytolysed. Loeb's key discovery was that, if eggs receiving transient treatment with one of these agents were then transferred to hypertonic sea water and finally to normal sea water, a large proportion of eggs developed to the larval stage. Some of the agents used by Loeb are listed in Table 3. To this list can be added treatment

TABLE 3. INDUCERS OF MEMBRANE ELEVATION AND PARTHENOGENETIC DEVELOPMENT IN SEA-URCHIN EGGS

No development occurs with 3; 1, 4, and 5 are the most efficient in terms of percentage of viable larvae produced.

1. Lipophils	Fatty acids, amines, organic solvents
2. Surfactants	Soap, saponin, bile salts
3. Extremes of tonicity	Very hypertonic salt, distilled water
4. Blood and tissue extracts	Mammalian blood only weakly active
5. Heterologous spermatozoa	Heated sperm active: heated homologous sperm inactive
6. Miscellaneous	Ammonia, silver ions

with a fluorescent dye and subsequent ultraviolet irradiation (Yamamoto, 1961). Loeb realized that one property was common to all activating factors: they were cytolytic or, in the case of biological fluids and cells, were capable of developing cytolytic properties. Loeb therefore proposed that the spermatozoon carries two fertilization factors: one that induces cytolysis of the cortical layer and a second "corrective factor" that interrupts cytolysis and permits the development of the egg.

Bataillon (1912) and Loeb performed similar experiments with frog eggs. These could be activated by pricking, but introduction into the egg of blood or other biological material was necessary in order that a large proportion of the eggs should develop into tadpoles. Parat (1933) found that acrosomes removed by microdissection from the large spermatids of the toad *Discoglossus* had a lytic action upon the surface membranes of the eggs. He showed in addition that the injection of one or two acrosomes caused a high proportion of *Discoglossus* eggs to develop parthenogenetically into larvae: a much higher proportion than could be obtained with other activating agents such as testis preparations and blood.

Tyler (1941) redefined the two activating factors on a cytological basis: the first initiates a phase which includes formation of the perivitelline space, rotation of orientation, and completion of the second meiotic division. The second initiates a regulatory phase involving normal cell division. Recently Pfohl (1970), studying frog eggs, investigated the distribution of the second or cleavage-initiating factor (CIF) in fractions obtained by homogenizing frog liver in sucrose-EDTA. The large granule fraction contained the bulk of the CIF, but an excess of this factor interfered with blastula development and also reduced the percentage of eggs that underwent first cleavage. Hoping to liberate CIF from the granules, Pfohl subjected them to ultrasonication and to freeze-thawing. Some acid phosphatase was released but the CIF was destroyed.

These results suggest that lysosomes are not involved in cleavage. But just as an excess of granule fraction interferes with the development of blastulae, it is possible that solubilized enzymes may have deleterious effects upon the egg in contrast to the limited localized stimuli that might originate from enzyme-carrying granules. In such experiments it must be borne in mind that the enzyme assayed (in this case p-nitrophenylphosphatase) may not be the appropriate marker. Dalcq (1963) demonstrated that yolk platelets stain for acid phosphatase and are similar in appearance to lysosomes. Brachet (Ref. VII) emphasizes that platelets play an active part in embryogenesis but suggests that phosphoprotein phosphatase and protease may be the significant enzymes in egg development. This implies the participation of lysosomal enzymes during the regulatory phase, while Loeb's transient cytolysis is consistent with the action of acrosomal hydrolases as triggers for the activation phase.

DISTRIBUTION OF ENZYMES IN SEA-URCHIN EGGS: EFFECT OF FERTILIZATION

In a study of proteolytic activities in eggs of *Paracentrotus lividus* Maggio (1957) prepared egg homogenates and separated them, at

20,000 *g*, into mitochondrial and cytoplasmic fractions. When comparable fractions from unfertilized and fertilized eggs were compared, it was clear that within as short a period as 5 minutes from fertilization there was a transfer of protease activity from the mitochondrial to the cytoplasmic fraction. Since Maggio's simple procedure would not have separated lysosomes from mitochondria, an equally valid conclusion would be that the immediate consequence of fertilization was either a release of lysosomal enzymes or a reduction in lysosome stability.

In the spring of 1969 Dr Allison and I availed ourselves of an invitation from Dr Alberto Monroy to study this question in the congenial atmosphere of his laboratory at Palermo. Our fractionation procedure (Fig. 5) yielded large granules (*G*), small granules (*M*), and a cytoplasmic fraction (*C*). We treated in this way both unfertilized eggs and eggs that had been fertilized 20 minutes before homogenization. Fractions *G* were further fractionated on a sucrose density gra-

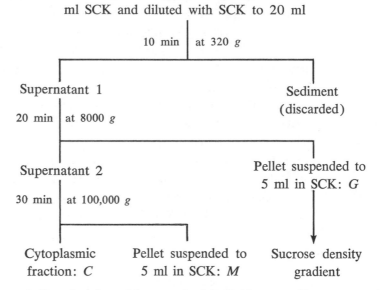

Figure 5. Fractionation of homogenized *P. lividus* eggs. Eggs were washed with sea water, sea water at pH 5 (to remove the jelly coat), and synthetic calcium-free sea water. SCK: 0.44 *M* sucrose containing 0.35 *M* KCl, 0.025 *M* citric acid, and NaOH to pH 6.8.

dient column (Fig. 6), and these fractions (I–V) were examined with an electron microscope. In addition they, as well as fractions M and C, were freeze-thawed and assayed for acid phosphatase activity at pH 4.5 and for protease activity with denatured haemoglobin at pH 7. (We established that two proteases were present with pH optima of 4.5 and 7.0. The latter enzyme was the more active and, unlike the former, required cysteine for full activity.) Enzyme determinations on one batch of eggs are given in Table 4. The protease results confirm Maggio's finding that activity is transferred, after fertilization, from particles (I–V, M) to the cytoplasmic fraction. The phosphatase assays reveal a considerable activation at fertilization. We cannot say at present whether this activation is a physiological phenomenon or an artifact arising from different responses of fertilized and unfertilized eggs to the fractionation procedure. However, the cytoplasmic enzyme/granule

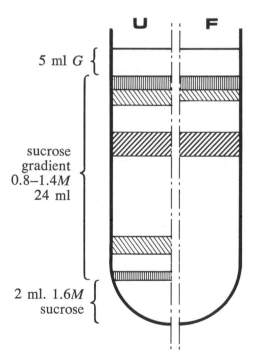

Figure 6. Separation on a sucrose-density gradient of the large granule fraction from unfertilized (U) and fertilized (F) *P. lividus* eggs.

TABLE 4. NEUTRAL PROTEASE AND ACID PHOSPHATASE ACTIVITIES IN
FRACTIONS OBTAINED FROM HOMOGENIZED UNFERTILIZED (U) AND
FERTILIZED (F) EGGS OF *Paracentrotus lividus*

I–V: density gradient fractions of large granules numbered from the top of
the gradient; *M*, "microsomes"; *C*, cytoplasmic fractions.

Frac-tion	Units/milligram protein				Distribution as % of total in U			
	Neutral Protease		Acid Phosphatase		Neutral Protease		Acid Phosphatase	
	U	F	U	F	U	F	U	F
I	3.7	1.80	8.2	9.3	30	18	45	64
II	9.6	3.77	4.9	5.9	15	12	5	30
III	4.0	2.97	3.9	2.9	3	3	2	5
IV	0.45		0.42		1		1	
V	0.53		0.03		2		<0.2	
M	0.56	0.30	0.49	0.71	2	1	1	2
C	0.16	0.27	0.23	0.78	47	67	46	128
					(100)	101	(100)	229

enzyme ratio was higher after fertilization. Grossman and Troll (1970)
have recorded a similar rapid transfer of membrane-bound hydrolases to
the cytoplasm when *Arbacia punctulata* eggs are fertilized.

Electron micrographs of fractions I–V and *M* from unfertilized
eggs all showed dense granules very similar to kidney lysosomes (Fig.
7). Fractions from near the top of the sucrose gradient column, and
especially fraction II, were richest in these granules. The presence of
the granules in fractions IV and V may have been due to their being
attached to other cell structures, or they may have become denser by
imbibing sucrose. Of the translucent vesicles in the granules some
(arrowed) appear to be undergoing extrusion as vesicles with double
membranes of which the outer one appears to be the granule membrane.
Although this result may be a sequel to the stresses of homogenization,
it seems that these stresses were not great enough to rupture the granule
membrane.

Michell, Karnovsky, and Karnovsky (1970) separated the cytoplasmic granules of guinea-pig polymorphonuclear leucocytes into six
groups by differential centrifugation. These showed considerable heterogeneity in terms of lysosomal enzymes, and acid phosphatase was concentrated mainly in the smallest and largest granules. In terms of specific

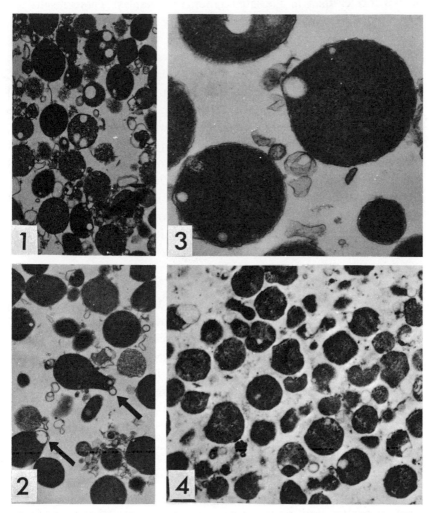

Figure 7. Electron micrographs of particle fractions from *P. lividus* eggs fixed with glutaraldehyde and postfixed with osmium tetroxide. *1*. Fraction II (high in protease; see Table 4) obtained from unfertilized *P. lividus* eggs (11000×). *2*. As *1*, showing extrusion of double-walled vesicles (11000×). *3*. As *1* (31500×). *4*. Kidney lysosomes separated from a mitochondrial fraction by sucrose density gradient centrifugation (11000×).

activities the acid phosphatase/protease ratio was 3–4 times higher in the former than in the latter group of granules. These results are comparable to those in columns 1 and 2 of Table 4, since fraction I, showing a high phosphatase/protease ratio, was also rich in very small vesicles. Michell et al. believe that their small particles, which are vesicles with a single limiting membrane, are not experimental artifacts. Though it is not possible to make a similar decision about the small vesicles in *P. lividus* eggs, it is known that acid phosphatase-positive particles in other invertebrate eggs have a wide range of sizes (Dalcq, 1963). A considerable particle heterogeneity and a predominance of acid phosphatase in the smallest particles also characterize homogenates of *A. punctulata* eggs (Jackson and Black, 1967).

Thus, although in general terms there is evidence that release of lysosomal-type enzymes is an immediate consequence of fertilization, a quantitative assessment of lysosomal function in biochemical terms is made difficult by the range of particle types.

LYSOSOMES AND MITOSIS

There appears to be no clear evidence for the involvement of egg lysosomes in mitosis. However, in transformed blood lymphocytes and in liver cells, mitosis is preceded by an increase in the lysosome population and by a movement of lysosomes towards the mitotic apparatus. Cell division is accompanied by a decrease in lysosome numbers and, in the case of regenerating liver, a release of hydrolytic enzymes (Adams, 1963; Kent et al., 1965; Allison, 1969).

INITIATION OF PROTEIN SYNTHESIS IN
SEA-URCHIN EGGS

Unfertilized sea-urchin eggs lack an effective system for protein synthesis, but incorporation of labelled amino acids can be detected very soon after fertilization or parthenogenetic activation (see reviews by Brachet, Ref. VII; Monroy, Ref. X; Monroy and Tyler, 1967).

Investigations in Monroy's laboratory (Maggio et al., 1964; Monroy, Maggio, and Rinaldi, 1965) have been directed towards determining the nature of the factor in *P. lividus* eggs that suppresses protein synthesis. The sudden onset of synthesis suggests activation of a link between ribosomes and mRNA, which were previously inaccessible to each other. Despite the absence of protein synthesis in unfertilized eggs, a low rate of amino acid incorporation occurred when RNA and ribosomes, both from unfertilized eggs, were brought together. The same RNA fraction induced a higher rate of incorporation by ribosomes from *P. lividus* blastulae or from rat liver. On the other hand, RNA from *P. lividus* embryos or from rat-liver nuclei promoted only low levels of incorporation by ribosomes from unfertilized eggs. Thus in the latter mRNA is functional, while ribosomes either are inaccessible to RNA or are in an inactive state. The former alternative appears less likely because the most active mRNA fractions from unfertilized eggs were the microsomal fractions containing the ribosomes. Incorporation at a low level, which takes place when both components come from unfertilized eggs, is presumably due to modifications of ribosomes during cell fractionation. When ribosomes from unfertilized eggs were treated with trypsin, they became able to incorporate amino acids actively in the presence of mRNA from the same source. The interpretation of these results (Monroy et al., 1965) is that mRNA synthesized during oogenesis becomes associated with ribosomes but that the complex is rendered nonfunctional by a protein coat. Proteases that are activated at fertilization would remove the coat and so initiate protein synthesis.

Working with eggs of *Hemicentrotus pulcherrinus*, Mano (1966) concluded that it was the mRNA of unfertilized eggs that was liberated from an association with protein by a protease activated at fertilization. Though the views of Monroy and Mano differ, the common theme is a sparking of the machinery for protein synthesis by proteolytic activity. Since fertilization of *P. lividus* eggs leads to release of protease from large granules and since the large-granule fractions are rich in lysosome-like organelles, the involvement of egg lysosomes in the initiation of protein synthesis becomes a reasonable working hypothesis.

HYDROLYTIC ENZYMES IN BACTERIA

In accordance with the view that bacteria and mitochondria have common ancestral origins, bacteria lack both lysosomes and membranous components in the cytoplasm analogous to the vacuolar digestive systems found in eukaryotic cells (Raven, 1970). However, they contain, and secrete, hydrolases comparable with those of lysosomes (Rogers, 1961; Pollock, 1962). Such enzymes may be membrane-bound, or they may occur in the space between the cell membrane and the cell wall. While discussing bacterial evolution, de Duve and Wattiaux (1966) suggest that selection of hydrolases in primitive bacterial forms was the vital step in the development of heterotrophic growth and the ability to reutilize biosynthetic material. A subsequent step would be development of heterophagy: a membrane-folding mechanism which would allow captured food and hydrolytic enzymes to be trapped together and which would also make larger areas of membrane available for metabolic exchange.

Although not related to bacterial reproduction, the release of penicillinase from bacteria has some interesting though tentative analogies with lysosomal function. Of the total penicillinase in *Bacillus licheniformis,* part is liberated spontaneously into the medium. Since bound enzyme is released from cells by trypsin, Pollock concluded that the mechanism for spontaneous release is also enzymatic (Pollock, 1961; Citri and Pollock, 1966). Ambler (1969) has compared amino acid sequences in the group of penicillinases that is secreted by *B. licheniformis* (NCTC 749) with the single penicillinase that is solubilized by treatment with trypsin. The "ragged" N-terminal sequences in the mixture of otherwise identical enzymes released by endogenous protease (Table 5) suggest that the latter enzyme has a wide specificity, or possibly that it is a mixture of proteases. (Similar work on another strain of *B. licheniformis* has indicated raggedness at the C-terminal end.) To account for the fact that proteolysis occurs only at the ends of the chain Ambler suggests that in its bound form penicillinase may be maintained in a compact, coiled type of configuration by being bound

TABLE 5. TERMINAL AMINO ACID SEQUENCES OF PENICILLINASES
RELEASED FROM *B. licheniformis* (749) BY ENDOGENOUS
PROTEASE AND BY TRYPSIN (Ambler, 1969)

Penicillinase-releasing factor	Sequence	
	N terminal	C terminal
Endogenous protease	LysThrGluMetLys—MetAsnGlyLys	
	ThrGluMetLys—MetAsnGlyLys	
	GluMetLys—MetAsnGlyLys	
Trypsin	ThrGluMetLys—MetAsnGlyLys	

at both termini to closely situated sites on a membrane. In such a situation only the terminal sequences would be freely accessible to proteolytic enzymes.

Application of these ideas to lysosomes deserves consideration, since it is now clear that some at least of the hydrolases in intact lysosomes are membrane-bound. Such binding provides a possible basis for the apparent lack of mutual digestion of enzymes within lysosomes that could be relevant to the distinction between the uncontrolled release of enzymes accompanying loss of membrane integrity at cell death and the controlled release called for in tissue remodelling, resorption, and secretion. When acid phosphatase from lysosomes was chromatographed by Beck et al. (1968), it separated into several active components. Other lysosomal enzymes behaved similarly. It would be interesting to know whether multiple forms of such enzymes were (i) isoenzymes built up from two types of subunit, as in the case of lactate dehydrogenase (Cahn et al., 1962); (ii) glycoproteins differing in carbohydrate composition (Barrett, 1969); or (iii) ragged variants on a common "ogen" amino acid sequence (Shamberger, 1969).

While there appears to be no evidence for the functioning of enzymes in sexually reproducing bacteria, one observation suggests that a hydrolase may be involved. When F^- K12 strains of *Escherichia coli* were mated with a large excess of certain strains of *Hfr* males, the great majority of the F^- cells were killed (Clowes, 1962). Subsequently it appeared (Clowes, 1969) that the death of the F^- cells was probably due to leakage of cytoplasm, which could be ascribed to imperfect repair of cell walls after transfer of DNA. Among the leaked products in a mating mixture was β-glucuronidase, which was absent from the culture media of the parent cells.

CONCLUSIONS

Brachet (Ref. VII) suggested that earlier theories of fertilization, especially with regard to responses of eggs to activation, could be reinterpreted on the basis of lysosomal functioning. I have attempted to classify what appears possibly to be relevant evidence for the involvement of lysosomes at the different stages of fertilization. However, decisions on whether the various items of evidence for lysosomal participation represent true control functions by lysosomal enzymes or merely responses by lysosomes to environmental changes brought about by as yet unidentified activators of gamete development must await further enquiries. I end on a note of caution with words of Louis Pasteur: *une fois engagé dans la voie de l'erreur, il est malaisé d'en sortir.*

ACKNOWLEDGMENTS

I am indebted to Dr A. C. Allison for Figure 1, *2;* to Dr H. Koenig for Figure 7, *4;* to Dr R. G. Saacke for Figure 4, *2;* and to Dr P. N. Srivastava for permission to reprint from his Ph.D. dissertation Figure 1, *1* and *3–5.* Figure 4, *3,* is reprinted from Bedford (1968) by permission of the author and of the Rockefeller University Press. I also wish to thank Mr. C. R. Brown for Figure 4, *1* and *4,* and Professor G. Millonig for Figure 7, *1–3,* which he prepared from egg fractions obtained by Dr Allison and me.

I am grateful to Drs R. P. Ambler and R. C. Clowes for information about their unpublished work. Finally I wish to thank Shell Research for a grant in support of my work at Palermo.

REFERENCES

I   Hayashi, T., ed. (1959). *Subcellular Particles,* New York, Ronald Press.

II   de Reuck, E. V. S., and M. P. Cameron, eds. (1963). *Ciba Foundation Symposium Lysosomes,* London, Churchill.

III Pathology Symposium: Lysosomes, *Fed. Proc. Amer. Socs. Exp. Biol.,* 23, 1009 (1964).

IV Campbell, P. N., ed. (1968). *Interaction of Drugs and Subcellular Components in Animal Cells,* London, Churchill.

V   Dingle, J. T., and H. B. Fell, eds. (1969). *Lysosomes in Biology and Pathology,* 2 vols., Amsterdam, North Holland.

VI Rothschild, Lord (1956). *Fertilization,* London, Methuen.

VII Brachet, J. (1960). *The Biochemistry of Development,* New York, Pergamon Press.

VIII Austin, C. R. (1961). *The Mammalian Egg,* Oxford, Blackwell Scientific Publications.

IX Mann, T. (1964). *Biochemistry of Semen and of the Male Reproductive Tract.* London, Methuen.

X   Monroy, A. (1965). *Chemistry and Physiology of Fertilization*, New York, Holt, Rinehart and Winston.

XI   Metz, C. B., and A. Monroy, eds. (1967). *Fertilization: Comparative Morphology, Biochemistry and Immunology*, 2 vols., New York, Academic Press.

Adams, R. L. P. (1963). *Biochem. J.*, 87, 532.

Allison, A. C. (1969). In V, vol. 2, p. 178.

Allison, A. C., and E. F. Hartree (1969a). *Biochem. J.*, 111, 35P.

Allison, A. C., and E. F. Hartree (1969b). In *Research in Reproduction*, edited by R. G. Edwards, vol. 1, No. 6, London, International Planned Parenthood Federation.

Allison, A. C., and E. F. Hartree (1970). *J. Reprod. Fert.*, 21, 502.

Allison, A. C., and M. R. Young (1964). *Life Sci.*, 3, 1407.

Allison, A. C., and M. R. Young (1969). In V, vol. 2, p. 600.

Ambler, R. P. (1969). Private communication.

Appelmans, F., R. Wattiaux, and C. de Duve (1955). *Biochem. J.*, 59, 438.

Aronson, N. N., and A. Davidson (1965). *J. Biol. Chem.*, 240, PC3222.

Austin, C. R. (1951). *Austr. J. Sci. Res.*, B, 4, 581.

Austin, C. R. (1952). *Nature*, 170, 326.

Austin, C. R. (1960). *J. Reprod. Fert.*, 1, 310.

Austin, C. R., and M. W. H. Bishop (1958). *Proc. Roy. Soc. (London)*, Ser. B, 149, 241.

Barrett, A. J. (1967). *Biochem. J.*, 104, 601.

Barrett, A. J. (1969). In V, vol. 2, p. 245.

Barrett, A. J., and J. T. Dingle (1967). *Biochem. J.*, 105, 20P.

Barros, C., J. M. Bedford, L. E. Franklin, and C. R. Austin (1967). *J. Cell. Biol.*, 34, C1.

Bataillon, E. (1912). *Ann. Sci. Nat. Zool.*, 16, 249.

Beck, C., S. Mahadevan, R. Brightwell, C. J. Dillard, and A. L. Tappel (1968). *Arch. Biochem. Biophys.*, 128, 369.

Bedford, J. M. (1968). *Amer. J. Anat.*, 123, 329.

Bedford, J. M. (1970). *Biol. Reprod.*, 2, Suppl., 128.

Bernstein, M. H., and R. J. Teichman (1968). *Anat. Rec.*, 160, 460; *J. Cell. Biol.*, 39, 14a.

Bishop, M. W. H., and J. Smiles (1957). *Nature*, 179, 307.

Blaschko, H., A. D. Smith, H. Winkler, H. van den Bosch, and L. L. M. van Deenen (1967). *Biochem. J.*, 103, 30C.

Bloom, G., and L. Nicander (1961). *Z. Zellforsch.*, 55, 833.

Boumer, J. M. W., and M. Gruber (1964). *Biochim. Biophys. Acta*, 89, 545.

Bowen, R. H. (1924). *Anat. Rec.*, 28, 1.

Burstone, M. S. (1961). *J. Histochem. Cytochem.*, 9, 146.

Cahn, R., N. O. Kaplan, L. Levine, and E. Zwilling (1962). *Science*, 136, 962.

Chang, M. C. (1951). *Nature*, 168, 697.

Citri, N., and M. R. Pollock (1966). *Adv. Enzymol.*, 28, 237.

Clermont, Y., R. E. Glegg, and C. P. Leblond (1955). *Exp. Cell Res.*, 8, 453.

Clowes, R. C. (1962). *Genet. Res.*, 4, 162.

Clowes, R. C. (1969). Private communication.

Colwin, L. H., and A. L. Colwin (1960). *J. Biophys. Biochem. Cytol.*, 7, 315.
Colwin, L. H., and A. L. Colwin (1967). In XI, vol. 1, p. 295.
Dalcq, A. M. (1963). In II, p. 226.
Dan, J. C. (1967). In XI, vol. 1, p. 327.
Daniel, J. C., and J. T. Millward (1969). *Exp. Cell Res.*, 54, 135.
Dott, H. M., and J. T. Dingle (1968). *Exp. Cell Res.*, 52, 523.
Duve, C. de (1963). *Sci. Amer.*, 208, 64.
Duve, C. de (1969). In V, vol. 1, p. 3.
Duve, C. de, B. C. Pressman, R. Gianetto, R. Wattiaux, and F. Appelmans (1955). *Biochem. J.*, 60, 604.
Duve, C. de, and R. Wattiaux (1966). *Ann. Rev. Physiol.*, 28, 435.
Epel, D., A. M. Weaver, A. V. Muchmore, and R. T. Schimke (1969). *Science*, 163, 294.
Ericsson, R. J. (1969). *Nature*, 221, 568.
Fawcett, D. (1970). *Biol. Reprod.*, 2, Suppl., 90.
Garcia-Bunuel, R., and D. Brandes (1966). *Amer. J. Obst. Gyn.*, 94, 1045.
Gomori, G. (1952). *Microscopic Histochemistry: Principles and Practice*, Chicago, University Press.
Grossman, A., and W. Troll (1970). *Biochim. Biophys. Acta*, 212, 192.
Hamner, C. E., and N. J. Sojka (1968). *Nature*, 220, 1042.
Hartree, E. F., and C. R. Brown (1970). Unpublished experiments.
Hartree, E. F., and P. N. Srivastava (1965). *J. Reprod. Fert.*, 9, 47.
Hathaway, R. R., and E. F. Hartree (1963). *J. Reprod. Fert.*, 5, 225.
Holt, S. J. (1963). In II, p. 114.
Howell, J. I., and J. A. Lucy (1969). *FEBS Letters*, 4, 147.
Jackson, C., and R. E. Black (1967). *Biol. Bull.*, 132, 1.
Kent, G., O. T. Minick, E. Orfei, F. I. Volini, and F. Madera-Orsini (1965). *Amer. J. Path.*, 46, 803.
Koenig, H. (1969). In V, vol. 2, p. 111.
Konechny, M. (1959). *Compt. rend. Soc. Biol.*, 153, 893.
Loeb, J. (1913). *Artificial Parthenogenesis and Fertilization*, Chicago, University Press.
Lucy, J. A. (1969). In V, vol. 2, p. 313.
McClean, D., and I. W. Rowlands (1942). *Nature*, 150, 627.
Maggio, R. (1957). *J. Cell. Comp. Physiol.*, 50, 135.
Maggio, R., M. L. Vittorelli, A. M. Rinaldi, and A. Monroy (1964). *Biochem. Biophys. Res. Comm.*, 15, 436.
Mancini, R. E., A. Alonso, J. Barquet, B. Alvarez, and M. Nemirovsky (1964). *J. Reprod. Fert.*, 8, 325.
Mano, Y. (1966). *Biochem. Biophys. Res. Comm.*, 25, 216.
Michell, R. H., M. J. Karnovsky, and M. L. Karnovsky (1970). *Biochem. J.*, 116, 207.
Miller, L. D., and D. T. Mayer (1960). *Res. Bull. Mo. Agric. Exp. Sta.* No. 742.
Monroy, A., R. Maggio, and A. M. Rinaldi (1965). *Proc. Nat. Acad. Sci. U.S.*, 54, 107.
Monroy, A., and A. Tyler (1967). In XI, p. 369.
Novikoff, A. B. (1959). In I, p. 1.

Novikoff, A. B., H. Beaufay, and C. de Duve (1956). *J. Biophys. Biochem. Cytol.*, **2**, Suppl., 179.

Novikoff, A. B., E. Essner, and N. Quintana (1964). In III, p. 1009.

Ono, Y., and S. Nojima (1969). *J. Biochem.* (Tokyo), **65**, 979.

Paigen, K. (1958). *J. Biol. Chem.*, **233**, 388.

Parat, M. (1933). *Compt. rend. Soc. Biol.*, **112**, 1131, 1134.

Pfohl, R. J. (1970). *Exp. Cell. Res.*, **61**, 433.

Pikó, L. (1967). *Int. J. Fert.*, **12**, 377.

Pincus, G. (1930). *Proc. Roy. Soc.* (*London*), Ser. B, **107**, 132.

Pollock, M. R. (1961). *J. Gen. Microbiol.*, **26**, 239, 255, 267.

Pollock, M. R. (1962). *The Bacteria*, edited by I. C. Gunsalus and R. Y. Stanier, vol. 4, p. 121, New York, Academic Press.

Raspé, G. (ed.) (1969). *Schering Symposium on Mechanisms Involved in Conception*, Braunschweig, Pergamon Press Vieweg.

Raven, P. H. (1970). *Science*, **169**, 641.

Reinauer, H., J. Brügelmann, W. Kurz, and S. Hollmann (1968). *Hoppe-Seyl. Z.*, **349**, 1191.

Revel, H. R., and E. Racker (1963). *Biochim. Biophys. Acta*, **43**, 465.

Rogers, H. J. (1961). In *The Bacteria*, edited by I. C. Gunsalus and R. Y. Stanier, vol. 2, p. 257, New York, Academic Press.

Ruffili, A. (1960). *Ric. scientifica*, **1**, 145.

Saacke, R. G., and J. O. Almquist (1964). *Amer. J. Anat.*, **115**, 143.

Shamberger, R. J. (1969). *Experientia*, **25**, 491.

Smith, A. U. (1951). In *Biochemistry of Fertilization and the Gametes* (Biochemical Society Symposium No. 7), edited by R. T. Williams, p. 3.

Srivastava, P. N., C. E. Adams, and E. F. Hartree (1965). *J. Reprod. Fert.*, **10**, 61.

Stambaugh, R., and J. Buckley (1969). *J. Reprod. Fert.*, **19**, 423.

Stambaugh, R., and J. Buckley (1970). *Soc. Study Reprod., Abstracts 3rd Ann. Meeting*, p. 7.

Tappel, A. L. (1969). In V, vol. 2, p. 207.

Toyoda, Y., and M. C. Chang (1968). *Nature*, **220**, 589.

Tyler, A. (1941). *Biol. Rev. Camb. Phil. Soc.*, **16**, 291.

Wada, S. K., J. R. Collier, and J. C. Dan (1956). *Exp. Cell Res.*, **10**, 168.

Wislocki, G. B. (1949). *Endocrinology*, **44**, 167.

Yamamoto, T. (1961). *Intern. Rev. Cytol.*, **12**, 361.

Yamamoto, T. (1962). *Embryologia*, **7**, 228.

Yamane, J. (1930). *Cytologia*, **1**, 394.

André Lwoff, perfectionniste

A. KIRN

C'est en tant qu'élève d'André Lwoff que je tente d'évoquer ici quelques-uns des souvenirs les plus marquants des relations scientifiques et humaines que j'ai entretenues avec lui depuis 1960. N'ayant pas collaboré à son oeuvre, ni travaillé dans son laboratoire, je ne peux pas parler de lui comme un élève le fait habituellement de son Maître. J'ai eu, cependant, la chance de profiter de son enseignement, de ses conseils, de ses encouragements et de bénéficier de son appui. André Lwoff m'a confié un travail pour lequel il a montré un grand intérêt et dont il a bien voulu guider les débuts, malgré la distance qui sépare Strasbourg de Paris. Le lecteur comprendra donc pourquoi les quelques souvenirs que je relate ici et qui ont trait essentiellement aux visites que je lui ai rendues et à la correspondance que nous avons échangée, se rapportent plus à sa personnalité qu'à son activité scientifique.

Ma première rencontre avec André Lwoff remonte à 1956, au "grand cours" de l'Institut Pasteur de Paris. Sacrifiant à une habitude pastorienne qui consiste à relever légèrement le col de la blouse, il se tient très droit derrière la paillasse qui fait office de chaire et parle d'un ton calme et mesuré, sans jamais hausser la voix. Dès les premières minutes de sa leçon sur la composition chimique des bactéries, l'attention de son auditoire lui est acquise. Chaque fait avancé, chaque point développé est solidement étayé par des résultats expérimentaux qu'il ne manque jamais de discuter. André Lwoff sait rendre sa leçon extraordinairement dynamique et vivante. C'est ainsi qu'il ne procède pas à un "dépeçage" de la bactérie suivi d'une analyse de ses différents constituants mais centre tout son exposé sur la biologie des protoplastes et s'efforce de démontrer les enseignements que l'on peut tirer de la physiologie bactérienne pour la compréhension de la chimie microbienne. L'étudiant participe à la leçon, car devant lui, avec lui, André Lwoff avec le sens de l'enseignement qu'il possède semble reproduire toutes les expériences qu'il décrit. Le cours terminé, il fait le tour des salles de travaux pratiques et devant de petits groupes d'étudiants, développe certains points particuliers qu'il n'a pas eu le temps d'approfondir. Il répond à toutes les questions et n'en esquive aucune. On retrouve dans chacune des leçons suivantes le même équilibre judicieux entre les faits acquis et les perspectives d'avenir et chaque exposé est également remarquable par la façon dont il est pensé, construit et présenté.

L'enseignement du "grand cours" est sanctionné par un examen comportant des épreuves écrites, pratiques et orales conduisant au diplôme de microbiologie. En tant que juge à l'oral de ce diplôme, André Lwoff a fait trembler de nombreuses promotions d'étudiants. Cette réputation était cependant largement exagérée. J'ai eu l'occasion de m'en rendre compte, le sort m'ayant désigné pour être interrogé par lui. J'avais lu son ouvrage *L'évolution physiologique* et j'eus la chance de me rappeler la définition des facteurs de croissance de Hopkins. Est-ce cela qui a rompu la glace? André Lwoff semble agréablement surpris de m'entendre citer Hopkins. Cela me met en confiance et je développe mon sujet. Il m'écoute attentivement mais ne manque pas de me poser de temps en temps quelques questions fort insidieuses, cherchant à savoir si ma lecture de *L'évolution physiologique* s'est bornée aux citations et aux têtes de chapitres. Rassuré sur ce point, il finit par me laisser partir avec une note fort honorable et une idée différente de sa sévérité d'examinateur.

Ce n'est qu'en 1960, après mon service militaire, que je reprends contact avec André Lwoff. L'image de la première visite que je lui ai faite à l'Institut Pasteur reste très vivante dans ma mémoire. Assis devant une hotte de culture de tissus, fabrication "maison", dans le laboratoire attenant à son bureau, le dos tourné à la fenêtre, les branches de ses lunettes réunies derrière la nuque par un tuyau en caoutchouc, il répartit des suspensions cellulaires dans des fioles de culture surmontées de capuchons argentés, le geste précis, efficace. Le visiteur est frappé par le nombre des bains-marie qui encombrent les paillasses du

laboratoire. Un échafaudage de tiges métalliques animées d'un mouve-
ment de va-et-vient communique ses oscillations à des fioles plongées
dans l'eau des bains-marie. Le clapotis qui en résulte se mêle au ron-
ronnement des moteurs, au sifflement des becs Bunsen et forme un
étrange bruit de fond. A cette époque, André Lwoff se penchait sur
le problème des facteurs non spécifiques de l'infection virale et en particulier sur l'action de la température. Il avait démontré que la thermosensibilité du développement du poliovirus est une caractéristique de
chaque souche, qu'elle varie avec les mutations et que celles-ci s'ac-
compagnent de variations du pouvoir pathogène. Ses expériences *in vivo*
montraient que l'hyperthermie expérimentale permet d'assurer la survie
des souris inoculées avec le virus de la poliomyélite ou celui de l'encé-
phalomyélocardite. Il en conclut que la fièvre, vis-à-vis de la maladie
virale devait être un moyen de défense naturel de l'organisme. Il me
suggère de le démontrer expérimentalement avec le virus vaccinal qui
produit une bronchopneumonie fébrile chez le lapin.

Périodiquement, je vais rendre compte à André Lwoff des résultats
obtenus et le renseigne sur les difficultés rencontrées. Ces visites ne sont
pas parmi mes meilleurs souvenirs malgré le sourire rassurant de "Mari-
nette" sa secrétaire qui me fait entrer dans la pièce mansardée où il
se tient assis à son bureau. André Lwoff m'écoute attentivement exposer
les résultats des dernières expériences et examine soigneusement les
tableaux et les graphiques que je lui soumets. Il se montre très exigeant
sur les protocoles expérimentaux et sur l'interprétation des résultats;
rien ne lui échappe et ses critiques souvent sévères sont cependant toujours justifiées. Exigeant sur le fond, il ne l'est pas moins sur la forme;
c'est ainsi qu'il est particulièrement allergique aux anglicismes qui truf-
fent certaines de mes phrases. Lorsqu'il m'entend prononcer des mots
comme one step, plaque, primer, translation, son sourire devient plus
ironique et il me prie de lui traduire ces termes en français.

Par ailleurs, André Lwoff se montre particulièrement féroce lorsque
le style employé dans la rédaction d'une communication scientifique
ne lui convient pas. J'ai eu l'occasion de subir les conséquences de ses
talents de censeur à plusieurs reprises. Les textes que je lui envoyais
pour avoir son avis me revenaient raturés, découpés, annotés de part
et d'autre et habituellement accompagnés de nombreuses remarques
aigres-douces sur le style. Ses critiques ne se limitent pas à l'usage de

la langue française. La réponse que j'ai reçue après lui avoir adressé mon premier manuscrit rédigé en anglais, en témoigne.

Mon Cher Kirn,

Votre note est fort intéressante mais vous écrivez l'anglais comme une vache espagnole. Je n'ai malheureusement pas le temps de corriger votre texte; peut-être trouverez-vous quelqu'un à Strasbourg?

Amicalement à vous,

André Lwoff

Il ne suffit pas d'élaborer des programmes de recherche encore faut-il pouvoir les réaliser, c'est-à-dire trouver le personnel et les moyens financiers nécessaires. André Lwoff ne se contente pas d'être un Maître à penser; il est pleinement conscient des difficultés qui se dressent sur le chemin de celui qui cherche à constituer une équipe et essaye de remédier à cette situation. Dans les commissions scientifiques des différents organismes de recherches français, il défend fermement la cause de la virologie. Au sein du conseil scientifique et en tant que président de la commission de microbiologie, il tente d'imprimer une nouvelle politique à l'INSERM.

Partisan résolu de grouper les hommes et les moyens, il lutte contre les saupoudrages des crédits de recherches trop souvent pratiqués auparavant. Il enseigne à ses élèves l'art et la manière d'obtenir des subventions et les aiguille vers les organismes aptes à satisfaire leurs demandes. "Il faut frapper à toutes les portes, il faut savoir mendier" me répète-t-il souvent. Ses leçons ont porté leurs fruits: il y a quelques semaines, il me confiait que, pour ce qui était de mendier, j'étais un de ses meilleurs élèves. Avant toute création d'un groupe de recherches de l'INSERM une commission d'enquête doit vérifier sur place si les conditions permettant l'implantation du groupe sont réunies. Ainsi en janvier 1967, André Lwoff se rend à Strasbourg accompagné par Charles Chany et par un fonctionnaire de l'INSERM pour étudier la possibilité de créer un groupe de virologie à l'Institut de Bactériologie de la Faculté de Médecine. La séance débute à 8 heures du matin. Pendant plus d'une heure, je suis sur la sellette et me trouve soumis à un feu roulant de questions sur le programme de recherches et les moyens en locaux et en personnel. Ensuite André Lwoff se rend dans le laboratoire où souffle un vent de panique: les chercheurs s'affairent autour des bains-marie et des centrifugeuses, les techniciennes vont et viennent, le nombre de pipettes cassées atteint ce jour-là un record encore jamais égalé. Il regarde travailler chacun de mes collaborateurs à tour de rôle puis pose quelques questions sur les techniques utilisées, les résultats obtenus et les perspectives de poursuite du travail. Si certains éprouvent alors quelques difficultés d'élocution, son extraordinaire sourire, la gentillesse qui se dégage de lui et les encouragements qu'on peut lire dans ses yeux ont vite fait de faire régner la confiance. Le rapport qui a sanctionné cette visite semble nous avoir été favorable puisque au mois d'avril de la même année le groupe a été créé.

La thermosensibilité du développement viral et l'application possible de cette notion à la thérapeutique des infections à virus n'a certes pas été le seul problème qui ait préoccupé André Lwoff au cours des dix dernières années. D'autres que moi sont plus aptes à parler des divers aspects de son oeuvre scientifique au cours de cette période et le feront sans doute.

L'évocation de ces quelques souvenirs n'a d'autre prétention que de montrer comment André Lwoff est apparu à l'un de ses élèves. Pour qui ne le connaît pas le portrait que j'ai esquissé de lui peut paraître un peu sévère. Qu'on ne s'y trompe pas, sa sévérité et son esprit critique ont été pour moi de puissants stimulants. Que ce soit en tant qu'enseignant, en tant que chercheur ou tout simplement en tant qu'homme, j'ai énormément appris au contact d'André Lwoff, sa personnalité et son dynamisme m'ont profondément marqué. Malheureusement il y a une chose qu'il n'a pas réussi à me communiquer: son admirable art d'écrire et j'imagine que ce n'est pas la lecture de ces quelques lignes qui l'incitera à me contredire.

Le modèle

PIERRE SCHAEFFER

*I love and honour this man, and the idea of "giving" him an article
as a memento of my regard was—and is—distasteful. . . . I wrote
it and feel that it is inadequate as a statement of my regard.*
PAUL BOHANNAN, in "Fest me no Schriften," *Science,* 1969

L'homme vrai n'a plus de moi. TCHOUANG TSEU

Un peu naïvement, je me suis proposé deux maîtres durant mes années
à l'Institut Pasteur: André Lwoff et Jacques Monod. Si je suis capable,
me disais-je, d'assimiler les réactions inspirées de l'un, les développements logiques de l'autre, quelle éducation j'aurai eue! Il y faudra sans
doute un effort d'analyse, mais comme dans les dissertations dont s'est
fortifiée ma jeunesse, la synthèse suivra d'elle-même, limpide et pé-
remptoire. Muni de ce *vade-mecum,* jusqu'où ne monterai-je pas? Procé-
dons avec méthode et commençons l'analyse.

André Lwoff—pour moi, "le patron"—a accédé à la Biologie avec
une formation médicale. D'une rigueur notoirement discutable, celle-ci
a le curieux mérite de vous conduire tout droit au seuil des plus grands
problèmes—et là de vous abandonner. Elle exalte avec raison, sous le
nom de sens clinique, l'étendue de la culture et l'intuition dans la recherche. L'oeuvre du patron est à cet égard médicale. Il s'est voulu
physiologiste cellulaire au sens strict. De la culture pure de protozoaires
en milieu défini à la définition fonctionnelle des vitamines, de l'évolution
régressive au prophage et à son induction, il dégage des concepts, qui
cessent du même coup de l'intéresser. Il les abandonne aux disciplines

voisines, Biochimie et Génétique, mieux armées pour en poursuivre l'étude, sans trop s'y aventurer lui-même ni susciter autour de lui le regroupement de biochimistes et généticiens, séduits par l'ampleur de ses vues. Père nourricier de ces disciplines, dont il se tient informé et connaît personnellement les représentants les plus illustres, il en évite cependant le contact direct et semble préférer la fière devise popularisée par Larousse: "Je sème à tous vents".

Et pourquoi s'imposer la rigueur contraignante de ces disciplines, quand le monde biologique autour de soi est plein de grands sujets pour qui sait les voir? Le patron y aurait perdu sa fantaisie créatrice, sa raison d'être. Il aime la forêt, et les arbres la cachent. Détachons-nous donc d'un sujet dont l'étude est désormais tracée, pour revenir à la Biologie Cellulaire, dont tant de domaines sont en friche. Si d'ailleurs la santé de la Biologie lui impose d'être conceptuelle, devenir moléculaire la tue. Gardons-nous donc de poursuivre un sujet au-delà des frontières de la discipline: un sujet est mort, vive le suivant. Voilà donc dégagée, sur le modèle lwoffien, ma première idée-force, celle du biologiste errant, chassé d'un sujet à l'autre par fidélité à sa discipline, qu'il sait la proie des disciplines voisines au moindre symptôme de rigueur scientifique.

Adopter ce modèle a des conséquences pratiques immédiates, concernant en particulier la taille de l'unité de recherche. L'errance est pleine de risques, que l'on ne saurait en conscience faire courir à de nombreux élèves. Elle s'accommode mal de ces rapports annuels dodus, dont les jeunes chercheurs ont l'exécrable habitude de faire dépendre leur subsistance même. On n'erre bien que seul, et titulaire. Quelques techniciennes, quelques distingués visiteurs mis à part, il faudra maintenir réduit l'effectif du groupe.

Une telle politique il est vrai a peu de chances d'être très populaire en période de contestation. L'hostilité qu'elle ne manquerait pas de soulever, dans l'Université notamment, pourrait la rendre inapplicable. En science comme ailleurs, le dogmatisme nous guette: quitte à en prendre à son aise avec le modèle, il ne s'agira pas de vouloir l'introduire de force dans un cadre qui n'est pas le sien.

Si Jacques Monod a été longtemps revendiqué par les physiologistes bactériens comme un de leurs maîtres, c'est à la suite d'un malentendu. L'appartenance était occasionnelle; affective sûrement pas, et moins encore exclusive. De la croissance bactérienne à la diauxie, à la

régulation des synthèses d'enzymes et à la théorie des transitions allostériques, nombreuses auront été les disciplines mobilisées pour l'étude du
même sujet: la régulation de l'expression du génôme. Mais sous ses
aspects techniques variables, l'expérience du jour découle toujours en
droite ligne de celles effectuées il y a dix, vingt et même trente ans.
(Années pendant lesquelles le monde, souvent à son insu, n'a guère
travaillé qu'à enrichir le dit sujet; et l'éléphant, pour survivre, n'a pu
mieux faire qu'imiter servilement le coli.) Les disciplines n'existent que
par l'idée préconçue qu'on en a, et les frontières entre elles, s'il en
subsiste, appellent de toute urgence le viol. La fidélité à la discipline est
une entrave, qui cesse d'exister dès qu'on la reconnaît comme telle. Par
quelle aberration sentimentale en ferions-nous une vertu scientifique?
Vis-à-vis du sujet, le problème est tout autre: il ne s'agit pas de fidélité,
mais du développement logique de son étude.

Comme la précédente, cette conception de la recherche impose au
laboratoire sa taille et sa composition. Multidisciplinaire, il ne peut être
que grand. Puisqu'il lui faut un titre, introuvable dans l'arsenal existant,
il faut en forger un. D'où le succès de l'expression "Biologie Moléculaire", dont j'ai entendu Jacques Monod donner, *cum grano salis,* une
définition qui on s'en doute fait grincer bien des dents: "ce qui m'intéresse".

Élève studieux, j'ai retenu la leçon et je tiens ma phrase lapidaire:
"Une discipline est morte, vive le sujet", ou encore: "Où j'ai choisi je
sème". (Quel éditeur progressiste l'adoptera, et comment illustrée?)
Mais je sens poindre une certaine inquiétude concernant ma synthèse
finale, sa limpidité et la sagesse de mon entreprise. Où est mon *vade-
mecum?* Quelle pirouette réconciliera mes deux modèles, pourtant irré-
cusables? Faut-il chercher dans l'errance (de sujet en sujet, de discipline
en discipline) leur plus grand commun diviseur? L'errance tout court
s'érige mal en précepte.

Un vertige me prend: l'analyse conduit-elle bien à la synthèse, ou
faut-il incriminer l'enseignement-secondaire-que-le-monde-nous-envie?
Est-il sage de construire le portrait-robot du parfait modèle, de s'exposer
à le voir galvaudé, polycopié, expédié par avion à ses innombrables
abonnés par une agence de diffusion scientifique? Mon talisman au départ se voulait plus personnel, requérait plus de mystère.

Il reste bien sûr la solution polythéiste, qui consisterait à conserver

les deux modèles côte à côte, comme ils ont vécu. Elle n'est pas dé-
pourvue d'avantages, qui pourraient apaiser mon sentiment d'échec. Un
dans chaque poche, j'aurais deux *"vade-meca";* selon mon humeur, ou
l'évolution de mon travail, je me justifierais rétrospectivement par l'un
ou par l'autre, préservant ainsi une liberté d'action que mes maîtres, eux,
avaient su conserver intacte. . . . Écrits au fil de la plume, ces mots
subitement me laissent rêveur, puis enfin m'illuminent: le voilà bien
l'insaisissable trait commun à mes deux modèles! Leur enseignement le
plus précieux était à découvrir dans *ce qu'ils n'ont pas fait!* Accepter
certaines contraintes, se donner des modèles, par exemple. . . .

    P.-S.- Cédant aux tentations du style dit alerte, j'ai parfois mis dans
la bouche de mes personnages des préceptes qu'ils n'ont jamais formulés.
J'en porte seul la responsabilité.

The Director, Institute of Cancer Research, Villejuif, 1969.